REVIEW OF INVERTEBRATE BIOLOGICAL CONTROL AGENTS INTRODUCED INTO EUROPE

Review of Invertebrate Biological Control Agents Introduced into Europe

Esther Gerber and Urs Schaffner

CABI Switzerland
Rue des Grillons 1
CH-2800 Delémont
Switzerland

CABI

CABI is a trading name of CAB International

CABI
Nosworthy Way
Wallingford
Oxfordshire OX10 8DE
UK

Tel: +44 (0)1491 832111
Fax: +44 (0)1491 833508
E-mail: info@cabi.org
Website: www.cabi.org

CABI
745 Atlantic Avenue
8th Floor
Boston, MA 02111
USA

Tel: +1 (617)682-9015
E-mail: cabi-nao@cabi.org

A catalogue record for this book is available from the British Library, London, UK.

Library of Congress Cataloging-in-Publication Data

Names: Gerber, Esther, author. | Schaffner, Urs, 1963- , author.
Title: Review of invertebrate biological control agents introduced into Europe / Esther Gerber and Urs Schaffner.
Description: Boston, MA : CABI, 2016. | Includes bibliographical references and index.
Identifiers: LCCN 2016031165 | ISBN 9781786390790 (hbk : alk. paper)
Subjects: LCSH: Insects as biological pest control agents--Europe. | Introduced insects--Europe. | Insect pests--Biological control--Europe.
Classification: LCC SB976.I56 G47 2016 | DDC 632/.96094--dc23 LC record available at https://lccn.loc.gov/2016031165

ISBN-13: 978 1 78639 079 0

Consignor: Federal Office for the Environment (FOEN), Soil and Biotechnology Division, CH-3003 Bern, Switzerland.
FOEN is an agency of the Federal Department of the Environment, Transport, Energy and Communications (DETEC).

Contractor: CABI Switzerland, Rue des Grillons 1, CH-2800 Delémont.

Authors: Esther Gerber and Urs Schaffner

Notice: This report was commissioned by the Federal Office for the Environment (FOEN). The contractor bears sole responsibility for the content.

Commissioning editor: Rachael Russell
Editorial assistant: Emma McCann
Production editor: Tracy Head

Typeset by SPi, Pondicherry, India.
Printed and bound in the UK by CPI Group (UK) Ltd, Croydon CR0 4YY.

Contents

Introduction and Summary

This review provides an overview of all documented releases of exotic (non-European) invertebrate biological control agents (IBCAs) into the environment in Europe and summarizes key information on the target species as well as on the biological control agent released. It is an update of *A Review of Biological Control in Western and Southern Europe* edited by Greathead (1976) and covers the period from 1897, when the beetle *Rodolia cardinalis* (Mulsant) was introduced into Portugal against the invasive cottony cushion scale, *Icerya purchasi* Maskell, until the end of 2009. This review is based largely on the BIOCAT database (Greathead and Greathead 1992), which contains records of the introduction of insect natural enemies, namely parasitoids and predators, for the control of insect pests worldwide. It does not include the introduction of other natural enemies for insect control, such as pathogens or nematodes. Also, it does not contain introductions into greenhouses and other protected cultivation where the agent is not expected to survive outdoors, although some of these organisms might subsequently also be found in nature. For instance, *Pseudaphycus flavidulus* (Brèthes), a wasp originating from Chile, has been introduced for studies in the laboratory, but subsequently has also been recorded in nature (Malausa *et al.* 2008). Most of the releases recorded in BIOCAT were made in the context of classical biological control (BC) projects, but in a few cases releases were also made in inundative BC projects. Developed by the late D.J. Greathead (former director of the International Institute of Biological Control, which is now part of CABI) and his wife, A.H. Greathead, BIOCAT was maintained continuously until the end of 2005. BIOCAT is currently being updated by CABI scientists. For this review, we used the database as it stood up to 2009, which included some 6000 records and the great majority of all insect introductions worldwide. Since the BIOCAT database is not yet accessible on the Internet, we refer to BIOCAT 2005 when citing information from the database.

In addition to BIOCAT, some key references on the introductions of arthropods into Europe were also considered (including Robinson and Hooper 1989; Orphanides 1996; Mifsud 1997; Noyes 2002; Lucchi *et al.* 2003; Jacas *et al.* 2006; Malausa *et al.* 2008; Rasplus *et al.* 2010; Roy and Migeon 2010). Arthropods used as IBCAs against invasive weeds were taken from Julien and Griffiths's (1998)

Biological Control of Weeds: A World Catalogue of Agents and their Target Weeds, fourth edition, which covers the period up to the end of 1996 and includes 1160 releases (including pathogens). Information on the recent release of an agent against Japanese knotweed in the UK was taken from Shaw *et al.* (2011). A review of additional sources of introductions of exotic organisms for classical biological control did not reveal any introductions of exotic nematodes or molluscs into Europe (Cock *et al.* 2010). Unattributed information on target species was taken from the Crop Protection Compendium (CABI 2007).

Following Greathead (1976), we delimited Europe to the set of countries belonging to the European Union (EU) and the European Free Trade Association (EFTA) (including Albania and states in former Yugoslavia); overseas territories were excluded.

BIOCAT 2005 also includes releases of insects native to some European countries that have been released in other European countries. These introductions are treated separately in Chapter 2, European Insect Biocontrol Agents Released in Europe.

Some introductions of BC agents into Europe have never been published. Therefore, this review may not provide the complete list of BC agents introduced into Europe. Nevertheless, the report includes a vast majority of the introductions, and hence provides a representative picture of the history of releases of exotic BC agents into the environment in Europe.

Species nomenclature follows the Global Biodiversity Information Facility (GBIF; www.gbif.org).

Countries of Origin and Destination

By far the largest number of introduced organisms are reported for Italy (Table 1), followed by France, Greece, Spain and Cyprus. In general, more organisms were introduced into Mediterranean countries than into countries of central and northern Europe.

Information on the source countries of the agents is not available for all introductions. From the data available, North America (Canada and USA) appears to be the most important source of exotic biocontrol agents released in Europe (Table 2).

Introduced biocontrol agents are frequently redistributed, i.e. populations from a country where the species has been established are then released into a third country. Some successful biocontrol agents have been spread literally worldwide (see, for example, *Encarsia berlesei* (Howard), Section 1.47.2), and for some their original native range is not known (Section 1.11.1). For Europe, the USA is the most important 'secondary donor country' (Table 3). In addition, both France and Italy have supplied other European countries with several exotic biocontrol agents (Table 3).

Several exotic BC agents have been reported from more countries than those in which they were introduced originally (e.g. Section 1.5.1; see also Table 4). This could be either because the species have spread naturally once released in Europe, as is the case for the Asian ladybird, *Harmonia axyridis* Pallas, or *Psyllaephagus pilosus* Noyes, a parasitoid of a pest on eucalyptus (Section 1.15.1), or because biocontrol

Table 1. Number of exotic biological control agents introduced into individual countries in Europe. Data are based on the species reports provided in this report.

Country	Number of agents
Italy	85
France	48
Greece	43
Spain	35
Cyprus	32
Former Yugoslavia	14
Former Czechoslovakia, Poland, UK	13
Portugal	8
Germany, Malta	6
Switzerland	5
Austria, Czech Republic	3
Belgium, Sweden	2
Croatia, Denmark, Hungary, Ireland, Macedonia, Netherlands, Slovenia	1

Table 2. Donor countries of the native range of biological control agents that were introduced into Europe. Data are based on the species reports provided in this report.

Country	Number of agents
USA (including Hawaii)	23
Canada	13
Eritrea	9
China, South Africa	5
India	4
Australia, Mexico, Japan	3
Algeria, Israel	2
Barbados, Brazil, Ethiopia, Libya, Morocco, Pakistan, Thailand, Tunisia, former USSR	1

agents are spread unintentionally by humans (Section 1.30.1). However, it is also likely that species have been introduced into more countries than have been reported in the literature available for this review.

European agents have also been moved within Europe (Chapter 2; see also Table 5). For instance, three European natural enemies were introduced against Moroccan locust, *Dociostaurus maroccanus* (THUNBERG), on Mediterranean islands (Section 2.3), and several species from Continental Europe were introduced to the UK against bark beetles (Section 2.3). In a single case, a Siberian predatory beetle, *Rhizophagus grandis* GYLLENHAL, was introduced into France to accelerate the natural spread of this species in Europe (Section 1.16.1).

Table 3. Secondary donor countries (i.e. not within the native range) acting as sources for releases of exotic biocontrol agents in European countries. Data are based on the species reports provided in this report.

Country	Number of agents
USA (including Hawaii)	21
France (including La Réunion)	13
Italy	6
Israel	5
India	4
Japan	3
Cyprus, Germany, South Africa, Portugal (including Madeira), former USSR	2
Belgium, Chile, Cuba, Egypt, Morocco, Switzerland, Uruguay	1

Targets

Exotic agents have been introduced against 59 specified pest species, and also against groups of insects such as aphids, scale insects and moths in general (Sections 1.55, 1.56 and 1.57). In addition, no specific target group is reported for one biocontrol agent (Section 1.59). Most pests targeted by exotic insect biocontrol agents are sap-sucking insects in the order Hemiptera (58%), predominantly aphids and scale insects (Fig. 1). Pests are further attributed to the orders Lepidoptera (22%), Coleoptera (10%), Diptera (5%) and Hymenoptera (3%).

European agents have been relocated within Europe against nine target species (Table 5). For four of them, i.e. *Adelges piceae* (Ratzeburg) (Section 2.1), *D. maroccanus* (Section 2.3), *Ips sexdentatus* (Boerner) (Section 2.5) and *Liriomyza trifolii* (Burgess in Comstock) (Section 2.6), biological control has been attempted using European agents only. It should be noted, though, that it is likely that relocations of European parasitoids or predators within Europe have not always been published. Hence, Table 5 is considered to provide only a subset of all relocations of European predators or parasitoids.

Some 73% (N = 125) of the agents introduced in European countries targeted pests on trees; 36 organisms were released for BC in timber and 87 in fruit production. Within the latter, mainly two crop types were targeted: 24% (N = 41) and 18% (N = 31) of all organisms introduced into European countries targeted pests on citrus and olive, respectively.

The number of biological agents introduced against pests of field crops is comparatively low. Only 16% of all agents were introduced for this purpose. Most of these agents (N = 17) were released against pests in potatoes (Section 1.29 and 1.43). Only one single species is reported to date to have been introduced specifically against a pest on ornamental trees (Section 1.1).

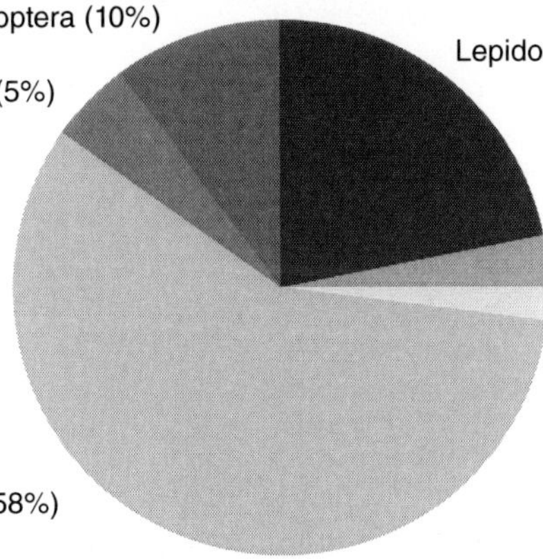

Fig. 1. Arthropod pests (*N* = 59) by order against which exotic biological control agents have been introduced in Europe.

A significant number of agents have been released against pests in more than one habitat, or pests attacking more than one commodity. For example, *Podisus maculiventris* (SAY) was released against Lepidoptera pests both in forests and in potato production (Sections 1.25.10 and 1.29.7). Moreover, several target species that are major pests in orchards or forests also attack ornamentals. Hence, pests of ornamental species have been targeted more often by arthropod biological control than indicated in Fig. 2. No information on the target habitat/crop was available for 17 organisms.

In the majority of cases, the considerable economic impact of the pest species was used as a legitimate reason for biocontrol agent introductions. However, biocontrol agent introductions are also reported against a citrus scale, *Unaspis yanonensis* (KUWANA) (Section 1.54), that does not occur in commercial citrus-growing areas in Europe and whose economic impact has been negligible (CABI 2007). In another case, releases of two exotic parasitoid species were apparently made for study reasons, i.e. to investigate the potential of introduced biocontrol organisms to adapt to new hosts (Starý 1995, 2002).

Biological control agents released

At least 176 insect species have been introduced as biocontrol agents in European countries, of which 165 have been identified to species level (Tables 4 and 5). Among these, parasitoids (134 species) were more numerous than predators (31 species), and organisms feeding inside the pest species (endophagous; *N* = 72) were more numerous than externally feeding ones (ectophagous; *N* = 56). Most introduced BC agents were wasps (Hymenoptera, 77%), predominantly species in the families Aphelinidae, Braconidae, Encyrtidae and Eulophidae. Beetles (Coleoptera, 17%), parasitic flies (Diptera, 4%), predatory bugs (Hemiptera, 2%)

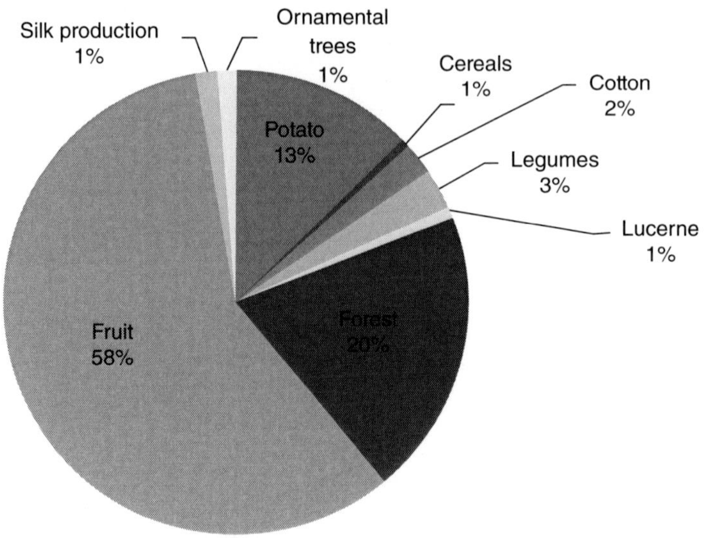

Fig. 2. Commodities targeted by arthropod biological control using exotic invertebrate biological control agents.

and one lacewing (Neuroptera) have also been introduced (Tables 4 and 5). At least 11 agents (6%) were introduced without identification at the species level (Table 4). The last introduction of an unidentified BC agent took place in the 1990s (Sections 1.38.2 and 1.44.5). In the case of *Metaphycus anneckei* Guerrieri and Noyes and *Metaphycus hageni* Daane and Caltagirone (Section 1.50.5), species were not identified correctly at the time they were released, and even now it is still not clear which species have been introduced into which country.

The first documented introduction of an exotic arthropod BC agent into Europe was in 1897 (Section 1.26). After a first period of increased BC activities, the number of releases dropped between the 1930s and the late 1950s, which could be explained by political unrest in the Mediterranean countries (e.g. Civil War in Spain) and the appearance of the first synthetic insecticides (Fig. 3). During the 1960s, the number of introductions of BC agents increased again, which coincided with the first appearance of books and articles documenting the adverse effects of the use of synthetic insecticides.

In the majority of cases, host specificity was not evaluated experimentally or studied in the field in the native range of the agents prior to their introduction (but see Section 1.16.1). The information on host specificity given in Tables 4 and 5 is based mainly on the information found in the Crop Protection Compendium (CABI 2007) and two additional databases (Noyes 2002; Yu *et al.* 2005), and is therefore likely to be incomplete. However, based on this information, 16 species can be considered as monophagous, i.e. only attacking the target pest. For 7 species, all other known hosts are in the same genus, and for 62 species, in the same family as the target pest. Another 46 species are recorded from other species in

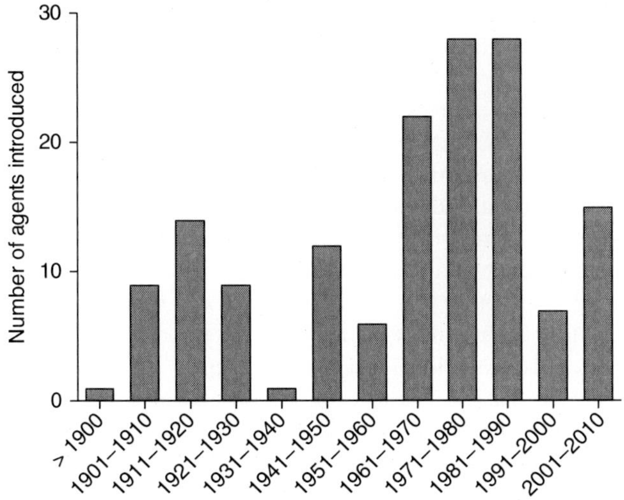

Fig. 3. Number of exotic species introduced into European countries per decade. Only first introductions are included; several species were introduced repeatedly into Europe.

the same order, while 25 species have an even broader host range, i.e. they attack species in more than one insect order.

Releases in Europe also include 'new associations'. The new association theory of biological control predicts that novel enemies that have not evolved with a target pest may be more effective in controlling the pest species than the co-evolved natural enemies (Hokkanen and Pimentel 1989). For instance, *Hyssopus thymus* Girault, a North American parasitoid, adopted a Lepidoptera pest of European origin in its exotic range. This parasitoid species was subsequently introduced and released into the native range of the pest (Section 1.49.1). Similarly, *Galeopsomyia fausta* La Salle, a parasitoid species from South America, was introduced because it adopted an Asian pest species in its native range that was also of concern to Europe (Section 1.44.4). In other cases, new associations might have been attempted, i.e. by releasing a predator beetle, known to feed mainly on aphids, against a Lepidoptera pest (*Ceratomegilla maculata*; Section 1.25.3). It should be noted, though, that new associations cannot be monophagous, and so non-target attack is a priori more likely.

Information on the non-target attack of released agents in Europe is scarce. Post-release studies conducted on the parasitoid *Aphidius colemani* Viereck, a species that was introduced against several pests in Europe (Sections 1.6, 1.19, 1.32), revealed that the species also attacked aphid species on non-cultivated plant species in the Czech Republic (Section 1.19.1).

While some species were released in small numbers only, others were released in very large numbers (e.g. *Encarsia perniciosi*; Section 1.18.2). In most cases, however, the number of released individuals either is not reported or is unknown (e.g. Section 1.19.1). In one case, parasitoids were released as infested hosts on twigs, which were then attached to branches in the field to release emerging adults (Section 1.47.2). Apart from the uncertainty about how many specimens have actually been released, this approach also includes the risk of releasing species other than the specific parasitoid intended.

The lack of pre-release studies (see above) also bears the risk of releasing biocontrol agents that are not adapted to the target pest because of morphological constraints (e.g. parasitoids with short ovipositors unable to reach target pest larvae, see Section 1.8), the requirement of alternative hosts, or because of poor synchronization with the pest species in the introduced range (e.g. Section 1.25.4).

Apparently, little attention has been given to interspecific competition between the agents released. For individual target species, up to 16 agents were released in Europe (Sections 1.8, 1.25, 1.43 and 1.50). Against *Bactrocera oleae*, 13 species were released in Italy alone (Section 1.8). In the case of the Lepidoptera pest *Prays oleae* on olives, at least five *Trichogramma* species were released in Greece that all occupied the same feeding niche, i.e. were egg parasitoids (Section 1.46).

1 Exotic Insect Biocontrol Agents Released in Europe

1.1 *Acizzia uncatoides* (Ferris and Klyver), Acacia Psyllid (Hem., Psyllidae)

Syn. *Psylla uncatoides* Ferris and Klyver

Acizzia uncatoides is native to Australia (Bellows *et al.* 1999). Its host range includes two Fabaceae genera, *Acacia* and *Albizia* (Ouvrard 2010). In 1954, it was first discovered on introduced *Acacia* trees in California (USA) (Nechols *et al.* 1995). In Europe, *A. uncatoides* was first reported in Italy in the mid-1970s. The species is now recorded from France, Italy, Malta, Montenegro, Portugal and Spain (Ouvrard 2010).

Feeding by *A. uncatoides* causes foliage chlorosis and tip dieback on new growth (Koehler *et al.* 1966). As with other Hemiptera, the honeydew produced by the psyllid is a medium on which sooty moulds can develop and become secondary pests.

Biological control has been applied successfully in California and Hawaii (USA). The list of natural enemies includes five predatory species.

1.1.1 *Harmonia conformis* (Boisduval) (Col., Coccinellidae)

Harmonia conformis is a predatory beetle native to Australia (Waterhouse and Sands 2001). The known host range also includes other Hemiptera, i.e. six Aphididae species, unspecified Pseudococcidae and two Chrysomelidae species (Coleoptera) (CABI 2007). *Harmonia conformis* has been introduced against *A. uncatoides* in Hawaii and mainland USA (BIOCAT 2005). It was further introduced into mainland USA against another psyllid, *Psylla pyricola*, and *Takecallis taiwanus* (Aphididae), and into Hawaii and New Zealand against an unspecified aphid species (BIOCAT 2005).

In 1998 and 2000, *H. conformis* originating from Australia was shipped to France and some 10,000 individuals were released in south-eastern France.

The species established and was also observed feeding on *Acizzia jamatonica* (KUWAYAMA) (Malausa *et al.* 2008).

1.2 *Acyrthosiphon pisum* HARRIS, Pea Aphid (Hem., Aphididae)

Acyrthosiphon pisum is probably of Palaearctic origin, but is now distributed virtually worldwide. It is recorded in Europe from all countries except Malta, Estonia, Lithuania, Slovenia, Slovak Republic and Liechtenstein, but probably also occurs in these as well.

Acyrthosiphon pisum is a species complex comprising several subspecies and biotypes that preferentially colonize different host plants. Major hosts in Europe include *Medicago sativa* L. (lucerne), *Pisum sativum* L. (pea), *Trifolium pratense* L. (purple clover), *Trifolium repens* L. (white clover) and *Vicia faba* L. (broad bean). The host range of *A. pisum* also includes wild species. Heavy infestations on peas can cause stunting, deformation, wilting and even death. Severe damage can occur due both to direct feeding and to virus spread. The species is a known vector of more than 30 plant virus diseases.

A range of insecticides (organophosphates, carbamates and pyrethroids) have been recommended for control of *A. pisum*. Further, a number of synthetic cultivars of lucerne have been released since the early 1990s, some with increased resistance to *A. pisum*.

Acyrthosiphon pisum has been reported as being attacked by native parasitoids. For example, *Aphidius ervi* and *Praon barbatum* have been found on *A. pisum* on lentil in Poland. Important predators of *A. pisum* occur in the families Coccinellidae, Syrphidae, Anthocoridae, Geocoridae and Chrysopidae. Birds may be significant natural enemies of this (relatively large) aphid species on sturdier legume species. The pest status of *A. pisum* can be further reduced by natural outbreaks of fungal diseases. Altogether, 34 parasitoid, ten pathogen and 73 predator species are recorded to attack *A. pisum*.

1.2.1 *Aphidius smithi* SHARMA AND SUBBA RAO, lucerne aphid parasite (Hym., Braconidae)

Aphidius smithi is a solitary endoparasitoid native to India and Pakistan (Fox *et al.* 1967). Known hosts include 33 genera and more than 70 species, all Aphidae (Yu *et al.* 2005). Besides Europe, it has been introduced into Argentina, Australia, Canada, Chile, Hawaii, Peru and the USA against the same target species (BIOCAT 2005). It was further released in New Zealand against another aphid pest, *Acyrthosiphon kondoi* SHINJI (BIOCAT 2005). In Europe, it was first released in Poland in 1960, and in Czechoslovakia in 1967. *Aphidius smithi* established and spread to other European countries. Today, it is recorded from Andorra, Albania, Austria, Belgium, Bulgaria, Croatia, Cyprus, the Czech Republic,

Denmark, Finland, France, Germany, Greece, Ireland, Italy, Lithuania, Moldova, the Netherlands, Norway, Poland, Portugal, Russia, Slovak Republic, Spain, Switzerland and the Ukraine (Yu *et al.* 2005; Rasplus *et al.* 2010).

No information on the impact of *A. smithi* on *A. pisum* was found in the literature.

1.3 *Aleurothrixus floccosus* Maskell, Woolly Whitefly (Hem., Aleyrodidae)

Aleurothrixus floccosus appears to be of Neotropical origin, because it is widely distributed in South America. Today, it is also recorded in Asia, Africa, Europe, North America and Oceania. In Europe, it was first reported in early 1966 from Spain and southern France, and later also from Cyprus, Italy, Greece, Malta, Portugal and the UK (Katsoyannos *et al.* 1997).

Aleurothrixus floccosus is a polyphagous species, attacking more than 20 different plant genera from various families. In the Mediterranean region, it almost exclusively infests species of the genus *Citrus*. *Aleurothrixus floccosus* affects the host plant by sucking sap from the infested young leaves. Sooty mould growing on honeydew deposits blocks light and air from the leaves, thereby reducing photosynthesis. High infestations can be detrimental, in particular to young plants.

The most important contributions to the control of *A. floccosus* have been through biological control programmes. In the Neotropical region, *A. floccosus* is controlled effectively by numerous naturally occurring parasitoids. Natural enemies include 11 parasitoid (mainly from two genera), nine predator and one pathogen species.

1.3.1 *Amitus spiniferus* (Brèthes) (Hym., Platygastridae)

Amitus spiniferus is a larval-adult parasitoid native to the Caribbean and Neotropical region. The only host recorded for this species is *A. floccosus* (Maskell) (CABI 2007). Besides Europe, it has also been introduced into Chile, Hawaii, la Réunion and the USA against *A. floccosus* (BIOCAT 2005).

In 1970, 350 *A. spiniferus* originating from Florida and Mexico were first released in southern Spain, where they became established (Greathead 1976). In 1973, *A. spiniferus* originating from Mexico was released in France (OPIE 1986; Malausa *et al.* 2008), but it is not certain whether the species has ever established there (CABI 2007). An undated release of *A. spiniferus* is reported for Italy (Sicily), but its establishment is considered uncertain (CABI 2007).

According to the information available, *A. spiniferus* seems always to have been co-released with *Cales noacki* Howard (see below), or released in regions where *C. noacki* has also been released. It is therefore difficult to assess to what extent *A. spiniferus* has contributed to the successful control of *A. floccosus*. In southern Spain, where *A. spiniferus* was co-released with *C. noacki*, successful control of *A. floccosus* was attributed entirely to *C. noacki* (Greathead 1976).

1.3.2 *Cales noacki* Howard (Hym., Aphelinidae)

Cales noacki is native to Central America (Katsoyannos *et al.* 1997). Known hosts include at least 12 other Aleyrodidae and Diaspididae species, as well as the moth *Phalera bucephala* (L.) (Notodontidae; Noyes 2002). Besides Europe, *C. noacki* was introduced into Hawaii, Kenya, Mexico, Morocco, Peru, la Réunion, Tunisia, Uganda and USA against the same target species (BIOCAT 2005).

In 1970, 100 *C. noacki* from Florida and Mexico were first released in southern Spain (Greathead 1976). In the same year, a culture of *C. noacki* was initiated in France, and in 1971, first releases of 400 individuals (100 females, 300 males) were made in an orchard in southern France (Onillon and Onillon 1972). Releases in Italy (Barbagallo *et al.* 1992) and Corsica (1973; OPIE 1986), Portugal (1978; Cavalloro and di Martino 1986), Sicily (1981; Cavalloro and di Martino 1986), Sardinia (1982; Ortu and Prota 1986; Barbagallo *et al.* 1992), Malta (1986; Mifsud 1997) and Greece (1992; Katsoyannis *et al.* 1997) followed. In Greece, inoculative releases were made four times with *C. noacki* obtained from Spain, and additional releases followed in 1993 (Katsoyannos *et al.* 1997). To date, the species is recorded from France, Greece, Italy and Malta, while its establishment in Portugal and Spain is considered uncertain (CABI 2007; Rasplus *et al.* 2010).

Substantial control is reported from all areas where *C. noacki* has been released in Europe (Katsoyannos *et al.* 1997). For instance, by the end of 1972, the density of eggs of *A. floccosus* had fallen to less than 0.5% of the 1971 density in the orchard in southern France into which *C. noacki* was originally released (Onillon 1973). *C. noacki* spread rapidly, and in early 1973 effective parasitism was recorded over 80 km^2 (Onillon and Onillon 1974).

1.3.3 *Eretmocerus paulistus* Hempel (Hym., Aphelinidae)

Eretmocerus paulistus is native to South America (Myartseva and Coronado-Blanco 2007). The only known host is *A. floccosus* (Noyes 2002). Besides Europe, it was also introduced into the USA against the same target species (Boardman 1977).

In 1970, 397 *E. paulistus*, originating from Florida or Mexico, were released in southern Spain. According to Greathead (1976), *E. paulistus* did not establish, but Rasplus *et al.* (2010) reported it as established in Albania.

1.4 *Aonidiella aurantii* (Maskell), California Red Scale (Hem., Diaspididae)

Aonidiella aurantii is considered to be indigenous to South-east Asia, but it has been introduced accidentally to many other countries worldwide. Its distribution includes most of the tropics, and it is also found in greenhouses in temperate areas. In Europe, it was first observed in 1960 in a few scattered locations in Italy. Today

A. aurantii is recorded from Cyprus, France, Greece (Crete), Italy (Sardinia, Sicily), Malta, Portugal (Madeira) and Spain (Canary Islands).

Aonidiella aurantii feeds on numerous host plants from various families, but is known mainly as an important pest on *Citrus* species. It feeds on a variety of plant parts, including leaves, branches, trunks and fruits. Heavy infestations set back, or even kill, newly planted trees and reduce fruit production.

Parathion was used very extensively from 1949 onwards, but one or more annual treatments were required and various pest repercussions resulted. High-pressure units for rinsing *Citrus* fruit in pack houses have been developed to remove *A. aurantii* and other armoured scale insects mechanically.

During recent years, there has been an increasing effort to keep California red scale under biological control, especially in California (USA), Australia, Israel and South Africa. Overall, 41 parasitoid, 45 predator and three pathogen species have been recorded as attacking *A. aurantii*.

1.4.1 *Aphytis holoxanthus* DeBach, circular black scale parasite (Hym., Aphelinidae)

Aphytis holoxanthus is a gregarious ectoparasitoid native to Asia. It was first described from material originating from Hong Kong (Katsoyannos *et al.* 1997). Known hosts include seven other Diaspididae species, *Aspidiotus nerii* Bouché, *Chrysomphalus aonidum* (L.), *Chrysomphalus dictyospermi* (Morgan), *Diaspis echinocacti* (Bouché), *Lepidosaphes beckii* (Newman), *Pinnaspis strachani* (Cooley) and *Selenaspidus articulatus* (Morgan) (Noyes 2002). Besides Europe, the species was also introduced into California, USA, against the same target species (BIOCAT 2005). It was further used as a biocontrol agent against another Diaspididae pest on *Citrus* spp., *C. aonidum*, and introduced into Argentina, Australia, Brazil, Egypt, Hawaii, Israel, Lebanon, Mexico, Peru, South Africa and USA (BIOCAT 2005).

Between 1960 and 1962, *A. holoxanthus* was released at several sites in Cyprus (Greathead 1976). The released material was reared from *A. holoxanthus* cultures received from Israel in 1959 and 1960 (Wood 1962). At the same time, three other *Aphytis* spp. (*A. coheni* De Bach, *A. lingnanensis* Compere and *A. melinus* DeBach) were released at the same sites against *A. aurantii*. *Aphytis holoxanthus* failed to establish; only single individuals were recorded during a short period after the releases (Greathead 1976). No additional releases are reported in the literature, but recently *A. holoxanthus* has been recorded in Belgium, the Czech Republic, Germany, Spain, France and the Netherlands (Rasplus *et al.* 2010).

1.4.2 *Aphytis lingnanensis* Compere (Hym., Aphelinidae)

Aphytis lingnanensis is a gregarious ectoparasitoid native to China, India and Pakistan (CABI 2007). Known hosts include *A. aurantii* and numerous other Diaspididae species (Noyes 2002). Besides Europe, the species has also been introduced into

Argentina, Australia, Chile, Israel, Mexico, Morocco, South Africa and the USA against *A. aurantii* (BIOCAT 2005). It was further used as a biological control agent against nine other Diaspididae *Citrus* pests and was introduced into Bermuda, Costa Rica, Cuba, Iran, Japan, Peru, the Solomon Islands and the former USSR (BIOCAT 2005; CABI 2007).

Between 1960 and 1962, *A. lingnanensis* originating from California, USA, were released at several sites in Cyprus, together with *A. melinus* (see below; Wood 1962). In 1962, 5650 individuals originating from the USA were released in Greece against both *A. aurantii* and *C. dictyospermi* (Section 1.11.2). An additional 14,000 individuals were released in Crete in 1963, but apparently the species failed to establish (Greathead 1976). In 1964, *A. lingnanensis* was also released in Sicily, Italy (Greathead 1976; Clausen 1978). Information about its establishment in Sicily is contradictory: according to BIOCAT 2005, it failed to establish, but Greathead (1976) reported establishment at one of the three sites in Sicily. To date, *A. lingnanensis* is recorded from Italy as well as from Albania, Cyprus, Greece and Spain (Noyes 2002; CABI 2007; Rasplus *et al.* 2010).

In most areas, *A. lingnanensis* has had a beneficial impact, either alone or along with other introduced *Aphytis* spp. (CABI 2007). While this species is considered to be particularly effective in coastal districts in California (De Bach *et al.* 1971), *A. melinus* (see below) has proved to be the most successful biological control agent in Greece, Italy and Turkey (Greathead 1976).

1.4.3 *Aphytis melinus* DeBach, golden chalcid (Hym., Aphelinidae)

Aphytis melinus is a gregarious ectoparasitoid native to India and Pakistan (CABI 2007). Its host range comprises numerous Diaspididae species, including many economically important ones (Noyes 2002), but also *Aphis nerii* Boyer de Fonscolombe (Hem., Aphididae) and *Scirtothrips citri* (Moulton) (Thys., Thripidae) (CABI 2007). Besides Europe, this parasitoid has also been introduced into Argentina, Australia, Chile, China, Israel, Morocco, South Africa and the USA against *A. aurantii* (BIOCAT 2005). It was further used in Iran, Peru and the former USSR as a biological control agent against five other Diaspididae *Citrus* pests (BIOCAT 2005; CABI 2007).

In 1961, *A. melinus* originating from California, USA, was first released in Cyprus (Wood 1962). In 1962, 13,000 individuals were released in Greece against both *A. aurantii* and *C. dictyospermi* (Morgan) (Section 1.11.3). *Aphytis melinus* spread quickly, established widely and largely displaced another introduced parasitoid against scale insects, *Aphytis chrysomphali* (Mercet) (Greathead 1976). In 1964, *A. melinus* originating from California was released at a site in Sicily, Italy (Greathead 1976), and additional releases were made in 1965 and 1967. In Europe, the species has also been introduced into France, Greece, Italy and Spain against other scale insects (Sections 1.7.1, 1.11.3). To date, *A. melinus* is recorded in Europe from Albania, Belgium, Cyprus, the Czech Republic, Denmark, France, Germany, Greece, Italy, the Netherlands, Portugal, Spain and Former Yugoslavia (Noyes 2002; CABI 2007; Rasplus *et al.* 2010).

Data collected in an International Organisation for Biological and Integrated Control (IOBC) cooperative project indicated that *A. melinus* was the most successful biological control agent against *A. aurantii* in Greece, Italy and Turkey (OILB 1970). Excellent control of both *A. aurantii* and *Saissetia oleae* (Olivier) has also been achieved using an integrated approach, by applying a single application of parathion and dimethoate in winter when *A. melinus* is inactive (Inserra 1968).

1.4.4 *Comperiella bifasciata* Howard (Hym., Encyrtidae)

Comperiella bifasciata is a solitary endoparasitoid native to eastern Asia (China; CABI 2007). Known hosts include at least 21 Diaspididae species (Noyes 2002). Besides Europe, the species has also been introduced into Argentina, Australia, Chile, Israel, Mexico, Swaziland, South Africa and the USA against *A. aurantii* (BIOCAT 2005). It was further used as a biological control agent against six other Diaspididae pests and introduced into Bermuda, Israel, Egypt, Philippines and New Zealand (BIOCAT 2005). In Europe, *C. bifasciata* was introduced into Cyprus, France and Greece against other scale insects (Diaspididae and Coccidae) (Sections 1.11.4, 1.56.6). To date, it is recorded as present in Belgium, Cyprus, the Czech Republic, France, Greece, Hungary, Italy, Moldova, the Netherlands, Romania, Russia, Spain, Ukraine and former Yugoslavia (Noyes 2002; Rasplus *et al.* 2010).

In 1969, *C. bifasciata* originating from USA was released in Greece (Greathead 1976). Since it failed to establish, a second release of individuals from Antibes, France, was made in 1972. The species became established and is also recorded as present in Greece (CABI 2007), but apparently has no major impact as a biological control agent. In 1972, *C. bifasciata* originating from the USA was released in France (Bénassy and Bianchi 1974), by which stage, however, it was already present at the Côte d'Azur. Its economic importance is negligible. Low levels of attack were recorded, presumably because it achieved three generations a year, while the host *A. aurantii* only passed through two (Bénassy and Bianchi 1974). Between 1985 and 1989, *C. bifasciata* was introduced into Cyprus, where it established but reached only low attack levels (3.8%; Orphanides 1996).

1.5 *Aphis spiraecola* Patch, Green Citrus Aphid (Hem., Aphididae)

Aphis spiraecola probably has its origin in the Far East, from where it has spread to many countries worldwide. As with many other aphids, it can be transported easily on fruits and ornamental plants to new areas. It is listed as being present in North America since at least 1907, and to date has a worldwide distribution in temperate and tropical regions. *Aphis spiraecola* was introduced into the Mediterranean region around 1939. To date, it is recorded from Bosnia and Herzegovina, Bulgaria, Croatia, Cyprus, France, Greece, Italy, the Former Yugoslav Republic of Macedonia, Malta, Montenegro, the Netherlands, Portugal, Serbia, Spain, Switzerland and the UK.

Aphis spiraecola is a moderately polyphagous species. Primary (winter) hosts are *Spiraea* and *Citrus* species; secondary host plants include species in over 20 families. It feeds on a variety of plant parts including leaves, inflorescence, stems and fruits. On severely affected fruit trees, the entire yield is at risk. Direct feeding is particularly damaging to young trees in spring. Sooty moulds, which thrive on aphid honeydew, contribute to the cosmetic damage of fruit, reducing its marketable value. The aphid can transmit at least ten virus species.

A number of insecticide regimes have been recommended to control *A. spiraecola* on its most important crop hosts; however, it has developed resistance to several insecticides.

Biological control has been attempted in Israel, the USA and the former USSR using five parasitoid species (all Aphelinidae, Hymenoptera) and three Coccinellidae (Coleoptera) species (BIOCAT 2005). A range of predators have been noted for *A. spiraecola*, mainly Chrysopidae, Coccinellidae and Syrphidae. *Harmonia axyridis* is the major natural enemy controlling aphid numbers in integrated pest management programmes in *Citrus* spp. in the Korean Republic. Additional natural enemies include two pathogen and 12 parasitoid species.

1.5.1 *Harmonia axyridis* Pallas, Asian ladybird (Col., Coccinellidae)

Harmonia axyridis is a polyphagous predator native to central and eastern Asia (CABI 2007). It is reported to be primarily a polyphagous arboreal species that inhabits orchards, forest stands and old-field vegetation. Known hosts include various aphid and scale species, but also immature stages of *Chrysoperla carnea* (Stephens) (Chrysopidae), *Danaus plexippus* L. (Nymphalidae) and eight species of Coccinellidae (CABI 2007). *Harmonia axyridis* has been used widely as a biological control agent of pest aphids and scale insects. It was favoured for biological control of aphids because of its large size, diverse dietary range, efficiency as a predator and wide niche colonization ability. Besides Europe, *H. axyridis* was also released in North America as early as 1916 (CABI 2007). In Europe, *H. axyridis* has been sold commercially as an augmentative biological control agent since 1982. Releases in the field are reported from Greece and France (Sections 1.52.1, 1.55.2). To date, it is recorded in various European countries, including Albania, Austria, Belarus, Belgium, Bulgaria, Croatia, the Czech Republic, Denmark, France, Germany, Greece, Hungary, Italy, Liechtenstein, Luxembourg, Former Yugoslav Republic of Macedonia, Montenegro, the Netherlands, Norway, Portugal, Romania, Russia, Serbia, Slovak Republic, Spain, Sweden, Switzerland, Ukraine and the UK (CABI 2007; Roy and Migeon 2010). With the establishment and rapid population build-up of *H. axyridis* in western Europe, significant and ongoing declines in the distribution of formerly common and widespread native ladybirds (e.g. *Adalia bipunctata* L.) have been reported (Roy *et al.* 2012).

A colony of *H. axyridis* was imported from France into Greece in 1993, and in 1994 adults were released in 11 *Citrus* orchards in four localities in Greece (Katsoyannos *et al.* 1997). In total, 620 adults were released at a density of 30–40 individuals/tree. In 1994, *H. axyridis* was recovered from seven localities (Katsoyannos *et al.* 1997), but it is probable that it established only temporarily (Brown *et al.* 2008).

1.5.2 *Lysiphlebus testaceipes* (Cresson) (Hym., Braconidae)

Lysiphlebus testaceipes is a solitary endoparasitoid of aphids native to North America (CABI 2007). The host range includes at least 151 species from 58 genera (Yu *et al.* 2005). Known hosts include various other *Aphis* species, *Melanaphis sacchari* (Zehntner), *Myzus persicae* Sulzer, *Rhopalosiphum maidis* Fitch, *Schizaphis graminum* (Rondani) and *Toxoptera aurantii* (Boyer de Fonscolombe) (Yu *et al.* 2005). Besides Europe, *L. testaceipes* has also been introduced into Argentina, Australia, Brazil, Burundi, Chile, China, Hawaii, India, Kenya, Morocco, Philippines, Tonga and the USA against other aphid pests (BIOCAT 2005; Yu *et al.* 2005). Similarly, in Europe, it was also released against other aphid pests, i.e. against *T. aurantii* in France, Italy and Spain (Section 1.52.2) and against *Aphis spiraephaga* Muller in the Czech Republic (Section 1.6.2). The species is further reported as being introduced into Portugal against unspecified aphids (Section 1.55.4). *Lysiphlebus testaceipes* established here successfully and is to date also recorded from Albania, Bosnia and Herzegovina, Bulgaria, Croatia, the Czech Republic, Denmark, France, Greece, Italy, Former Yugoslav Republic of Macedonia, Montenegro, Portugal, Serbia and Spain (Yu *et al.* 2005; Rasplus *et al.* 2010).

Between 1973 and 1974, *L. testaceipes* was imported from Cuba to Mediterranean France for the biological control of *A. spiraecola* and other *Citrus* aphids (Starý *et al.* 1988). The founder population for this introduction was reared from *A. nerii* in Cuba and shipped to former Czechoslovakia, where it was reared on *Aphis craccivora* L. and *Aphis fabae* Scop. (Starý *et al.* 1988). Some 2300 mummies (parasitized aphids) were sent to France in 1973, followed by another 3000 mummies in 1974, to establish a rearing. In 1973, 1000 parasitoids were released in southern France and Corsica, and in 1974 an additional 1000 individuals were released in southern France and 2000 in Corsica. At the individual release sites, several dozens to 300 individuals of either parasitized live aphids, mummies and/or adults were released, using different release methods (confined, semi-confined and open; Starý *et al.* 1988).

Overall, *L. testaceipes* is considered to be a relatively successful agent against aphids (Starý *et al.* 1988) and to have some impact on pest numbers of *A. spiraecola* (CABI 2007). However, the parasitoid cannot complete its development on *A. spiraecola*; parasitized aphids either die or stop producing offspring, and no further parasitoids are produced from mummies of this relatively small aphid species (CABI 2007).

1.6 *Aphis spiraephaga* Muller, Spireas Aphid (Hem., Aphididae)

Aphis spiraephaga was of Central Asian origin (Starý 1995), from where it dispersed westwards and was first recorded in Czechoslovakia in 1956 (Starý 1995). The species has become widely distributed in most of Europe and parts of Asia Minor, and is now recorded from Austria, Bulgaria, the Czech Republic, Denmark, Germany, the Netherlands, Romania, Slovak Republic and Sweden, and from the European part of Russia (Starý 1995).

Aphis spiraephaga is associated mainly with various ornamental *Spiraea* shrub species (Starý 1995). Further secondary hosts in the Czech Republic include *Bellis perennis* L. (Asteraceae), *Carum carvi* L. (Apiaceae), *Arabis hirsuta* L. (Brassicaeae) and *Valeriana officinalis* L. (Valerianaceae). Occasionally, *A. spiraephaga* can damage ornamental plants; honeydew production covers the leaves with a sticky layer of black mould. However, the releases of two exotic parasitoid species (see below) appear mainly to have been motivated not by economic interests but by scientific interest in studying the potential of introduced biocontrol organisms to adapt to new hosts (Starý 1995, 2002).

Biopesticides and metabolites from autochthonous plants were tested for the control of *A. spiraephaga* (Brudea 2009). The efficacy of the active ingredients spinosad, azadirachtin and milbecmectin reached 95–100% after 4 days.

Only one parasitoid species is recorded in the literature on *A. spiraephaga* from its original range in Central Asia, while at least 23 species of predators and ten parasitoids are recorded from the Czech Republic (Starý 1995).

1.6.1 *Aphidius colemani* Viereck (Hym., Braconidae)

Aphidius colemani is a solitary endoparasitoid of aphids that was initially thought to originate from India, but reclassification of two other species as junior synonyms of *A. colemani* extended its geographical range from Central Asia to the Mediterranean (CABI 2007). This endoparasitoid is recorded from 76 Aphididae host species from 32 genera (Yu *et al.* 2005). Primary hosts are *M. persicae*, *M. nicotianae* Blackman and *Aphis gossypii* Glover, against which it has been particularly effective as a biological control agent (CABI 2007). *Aphidius colemani* is used mainly as a commercial biocontrol agent in glasshouses, targeting primarily *M. persicae.* The species is likely to have escaped from greenhouses; for instance, it accidentally escaped and subsequently established in cereal fields on *Rhopalosiphum padi* (L.) in Germany (Starý 2002). Besides Europe, it has been introduced against other aphid species in Africa, Australia, Brazil, Central America, Tonga and the USA (BIOCAT 2005; CABI 2007). In Europe, *A. colemani* was also introduced against *Diuraphis noxia* (Kurdjumov) and *Melanaphis donacis* (Passerini) (Sections 1.19.1, 1.32.1). It is considered established in Albania, Austria, Belgium, the Czech Republic, Denmark, Finland, France, Germany, Greece, Hungary, Ireland,

Italy, Lithuania, Malta, the Netherlands, Norway, Poland, Portugal, the Slovak Republic, Spain, Sweden, Switzerland and the UK (Yu *et al.* 2005; CABI 2007; Rasplus *et al.* 2010).

Aphidius colemani was introduced into the Czech Republic from Chile and reared on *A. fabae* Scopoli (Starý 1995). Together with another simultaneously released parasitoid species, *L. testaceipes* (Section 1.5.2), 7000 individuals were released in 1992, and another 8000 individuals in 1993 (Starý 1995). The establishment of *A. colemani* is considered as positive in supplementing the native parasitoid guilds, as well as in contributing to the overall biodiversity of hymenopterous parasitoids in the cultivated landscape (Starý 2002).

1.6.2 *Lysiphlebus testaceipes* (Cresson) (Hym., Braconidae)

Lysiphlebus testaceipes is a solitary endoparasitoid of aphids native to North America (CABI 2007). Its host range includes at least 151 host species from 58 genera, all Aphidae (many *Aphis* species) (Yu *et al.* 2005). Besides Europe, the species was also introduced into Argentina, Australia, Brazil, Burundi, Chile, China (Shaanxi), India, Kenya, Morocco, Philippines, Tonga and the USA (Hawaii) against other aphid pests (BIOCAT 2005; Yu *et al.* 2005). It has been further recorded from Mongolia and Central and South America, where it was released against other aphid species, including *A. spiraecola* and *T. aurantii* (Sections 1.5.2, 1.52.2, 1.55.4). To date, *L. testaceipes* is recorded from Albania, Bosnia and Herzegovina, Bulgaria, Croatia, Czech Republic, Denmark, France, Greece, Italy, Former Yugoslav Republic of Macedonia, Montenegro, Portugal, Serbia and Spain (Yu *et al.* 2005; Rasplus *et al.* 2010). *Lysiphlebus testaceipes* was imported together with *A. colemani* from Chile to the Czech Republic and reared on *A. fabae* (Starý 1995). In 1992, 7000 individuals were released, followed by 8000 in 1993 (Starý 1995). Permanent establishment of this species is reported from the Czech Republic (Starý 1995).

1.7 *Aspidiotus nerii* Bouche, Aucuba Scale (Hem., Diaspididae)

Aspidiotus nerii is considered to be native to the Mediterranean area. Today, it has a worldwide distribution, due primarily to the transport of infested plant material by humans. In Europe, it has been recorded from Bosnia and Herzegovina, Bulgaria, Croatia, Cyprus, France, Greece, Italy, Former Yugoslav Republic of Macedonia, Malta, Montenegro, Poland, Portugal, Romania, Russia, Serbia, Spain and the UK.

Aspidiotus nerii is a highly polyphagous insect that has been recorded on hundreds of host species in over 100 plant families. Its many hosts include agricultural crops, palms, cut flowers and woody ornamentals. Major hosts include *Actinidia chinensis* Planchon (Chinese gooseberry), *Albizia julibrissin* Durazzini (silk tree),

Citrus and *Olea* (olive). *Aspidiotus nerii* feeds on a variety of plant parts, including seedlings, flowers, fruits, leaves and stems. Usually, *A. nerii* is considered only a minor pest on most of its hosts. However, it is particularly important where the aesthetic value of the crop is high; for example, in cut flowers and ornamentals. In olive crops, the presence of a single scale makes the fruit unmarketable.

Pruning and allowing adequate spacing between plants throughout cultivation can reduce the spread of infested material. Mechanical control can be achieved by scraping and scrubbing to remove scales, and chemical control is usually carried out by spraying with horticultural oils (mineral oils) and organophosphorus insecticides at critical points during the season. *Aspidiotus nerii* populations can be controlled successfully by natural enemies. Introductions of parasitoids for biological control are reported from Peru and Israel (BIOCAT 2005). Overall, 17 parasitoid and nine predator species are recorded on *A. nerii.*

1.7.1 *Aphytis melinus* DeBach (Hym., Aphelinidae)

Aphytis melinus, a gregarious ectoparasitoid mainly of scale insects, is native to India and Pakistan (CABI 2007). Known hosts include at least 12 other Diaspididae species including many economically important species (Noyes 2002), but also *A. nerii* (Aphidae, Hemiptera) and *S. citri* (Moulton) (Thripidae, Thysanoptera) (CABI 2007). *Aphytis melinus* was first introduced into California (USA) from India and Pakistan in the 1950s against *A. aurantii.* It established successfully and was subsequently imported into many parts of the world. Besides Europe, the species was also introduced into Peru against *A. nerii* and into Argentina, Australia, Chile, China, Iran, Israel, Morocco, Peru, South Africa, Turkey and the USA against other Diaspididae pest in *Citrus* (BIOCAT 2005; CABI 2007). In most areas, *A. melinus* has had a beneficial impact, either alone or along with other introduced *Aphytis* spp. (CABI 2007). In Europe, *A. melinus* has also been introduced to control two other scale insects in Cyprus, France, Italy and Spain (Sections 1.4.3, 1.11.3). *Aphytis melinus* is recorded in Europe from Albania, Belgium, Cyprus, the Czech Republic, Denmark, France, Germany, Greece, Italy, the Netherlands, Portugal, Spain and former Yugoslavia (Noyes 2002; CABI 2007; Rasplus *et al.* 2010).

In 1976, *A. melinus* was released in Crete, Greece, against *A. nerii*, with some reported impact on pest numbers (OPIE 1986; CABI 2007).

1.8 *Bactrocera oleae* (Rossi), Olive Fruit Fly (Dip., Tephritidae)

Syn.: *Dacus oleae* Gmelin

Bactrocera oleae is considered to be native to Africa (Nardi *et al.* 2005), but its current distribution extends throughout the olive-growing zone worldwide, including many Asian countries, as well as California (USA) and Mexico (Nardi *et al.* 2005).

In Europe, it is recorded as present in Albania, Croatia, Cyprus, France, Greece, Italy, Malta, Portugal, Serbia, Slovenia, Spain and Switzerland.

In Europe and Africa, *B. oleae* has a narrow host range, attacking only wild and cultivated olive fruits (*Olea* spp.). In California, trees of the families Rosaceae, Rutaceae, Anacardiaceae, Fabaceae, Lythraceae and Malpighiaceae are also infested, but olives are the preferred host. *Bactrocera oleae* is the major insect pest of olive crops worldwide, and is responsible for losses of olive oil value and table olives. Oil made from infested fruits is up to 12 times as acidic as normal.

Keeping orchard areas clean of fallen fruit is a common sanitary practice that prevents many larvae from being able to emerge and pupate. Cover sprays of entire crops are sometimes used, but the use of bait sprays is more economic and more environmentally acceptable. A bait spray consists of a suitable insecticide mixed with protein bait, since both males and females are attracted to protein sources.

Attempts to control *B. oleae* in the Mediterranean using biological control started as early as 1911. Overall, 26 parasitoid, four predator and 16 pathogen species are reported from *B. oleae*, of which 15 parasitoids and one predator have been released for biological control of *B. oleae* in Europe.

1.8.1 *Belonuchus rufipennis* (Fabricius) (Col., Staphylinidae)

The native range of the predatory beetle *Belonuchus rufipennis* extends from north-eastern USA to Argentina (Frank and Barreray 2010). Known hosts include Tephritidae and Drosophilidae (both Diptera; Greathead 1976), but also the sap-feeding beetle species *Carpophilus humeralis* (Fabricius) (Col., Nitidulidae), against which it was released as a biocontrol agent in Hawaii (BIOCAT 2005). In Europe, *B. rufipennis* was also released against *Ceratitis capitata* (Wiedemann) in Italy (Section 1.9.2).

Between 1939 and 1941, *B. rufipennis* was released against *B. oleae* in Italy and established. However, no significant impact on the population density of the target species was observed (Russo 1959).

1.8.2 *Biosteres longicaudatus* Ashmead, longtailed fruit fly parasite (Hym., Braconidae)

Syn.: *Opius longicaudatus* Ashmead; *Diachasmimorpha longicaudata* Ashmead
Biosteres longicaudatus is a solitary endoparasitoid, reported as native to Taiwan and Thailand (CABI 2007). At least 31 host species are recorded from 13 genera, mainly Tephritidae species (including species in the genera *Anastrepha*, *Bactrocera*, *Biosteres*, *Dacus* and *Ceratitis*), but also species from other families and orders, i.e. Aphidae (Hemiptera), Noctuidae, Erebidae, Pyralidae/Crambidae and Tortricidae (Lepidoptera) and Agromyzidae (Yu *et al.* 2005). Besides Europe, *B. longicaudatus* was also introduced into Argentina, Australia, Belize, Cape Verde Islands, Costa Rica, Dominica, Ecuador, Fiji, Israel, Madagascar, Marianas, Mauritius, Mexico,

Nicaragua, Peru, Philippines, la Réunion, Saint Kitts and Nevis, Trinidad and Tobago, the USA (Hawaii and Florida) and Zambia against 12 other Tephritidae species (BIOCAT 2005; CABI 2007).

In 1952, 'a few individuals' of *B. longicaudatus* collected in Hawaii were released in Greece, but none survived (Pelekassis 1974). An additional 1500 individuals were released in 1966, but this number also included *Fopius arisanus* and *Diachasmimorpha tryoni* (Section 1.8.11; Clausen 1978). According to H.G. Stavraki (cited in Greathead 1976), while no *B. longicaudatus* have been recovered from these releases, it is now reported as established in Greece (CABI 2007). No information on its impact on the target species was found.

1.8.3 *Bracon celer* Szépligeti (Hym., Braconidae)

Bracon celer is a solitary, ectoparasitic idiobiont native to Ethiopia, Kenya and South Africa (WaspWeb 2004–2010). It has been reported as the most abundant parasitoid attacking *B. oleae* in South Africa and Kenya, with parasitism rates of up to 87% in South African olive orchards (references in Sime *et al.* 2006). Known field hosts also include *C. capitata* and *Trirhithrum nigrum* (Graham) (both Tephritidae, Diptera), and in the laboratory it has also developed in *Parafreutreta regalis* Munro, a candidate weed biological control agent for Cape-ivy (*Delairea odorata* Lamaire) in North America (Sime *et al.* 2006; Nadel *et al.* 2009).

In 1914, *B. celer* was collected in Eritrea and in 1914/15, 155 individuals were released in Italy (Clausen 1978), where the species established successfully (Yu *et al.* 2005). *Bracon celer* from South Africa was also released in Greece, but failed to establish there (Robinson and Hooper 1989).

According to Greathead (1971), *B. celer* is unsuccessful as a biological control agent because Italian olive skins are thick and only parasitoids with long ovipositors can parasitize easily olive fruit flies feeding inside the fruits.

1.8.4 *Cirrospilus variegatus* (Masi) (Hym., Eulophidae)

Syn.: *Zagrammosoma variegatum* Masi

Cirrospilus variegatus is an ectoparasitoid native to the Afrotropical region (Ethiopia, Kenya, South Africa, Tanzania, Uganda) (WaspWeb 2004–2010). Known hosts include mainly leafminers in Diptera (Tephritidae and Agromyzidae) and Lepidoptera (including *Cameraria ohridella* Deschka and Dimić and species in the genera *Phyllonorycter*, *Phyllocnistis*, *Stigmella*, *Leucoptera* and *Lyonetia* (CABI 2007; references in Yefremova 2008)). In Europe, the species is reported from Hungary, Italy, Montenegro, Serbia, Slovak Republic, Sweden and Switzerland (Noyes 2002; CABI 2007).

In 1914, *C. variegatus* was collected in Eritrea, and in 1914/15 some 500 individuals were released in Italy (Clausen 1978), where it established successfully (CABI 2007). However, the species is unsuccessful as biological control agent, probably because its ovipositor is also poorly adapted to parasitize its host through the thick skin of Italian olives (Greathead 1971).

1.8.5 *Euderus cavasolae* (Silvestri) (Hym., Eulophidae)

Euderus cavasolae is an ectoparasitoid native to Eritrea (WaspWeb 2004–2010). No other hosts are reported for this species (Noyes 2002).

In 1914, *E. cavasolae* was collected in Eritrea, and in 1914/15, 1200 individuals were released in Italy (Clausen 1978), where it established (CABI 2007). As with the other species introduced in Italy (see above and below), Greathead (1971) argues that *E. cavasolae* is unsuccessful as a biological control agent because Italian olive skins are thick and only parasitoids with long ovipositors can parasitize *B. oleae* readily inside these fruits.

1.8.6 *Eupelmus afer* Silvestri (Hym., Eupelmidae)

Eupelmus afer is an ectoparasitoid native to Africa. It has been reared from *B. oleae* feeding on cultivated and wild olives in South Africa (Wharton and Yoder 2016). No other host is reported for this species (CABI 2007).

In 1914, *E. afer* was collected in Eritrea, and in 1914/15, 179 individuals were released in Italy (Clausen 1978). According to CABI (2007), the species has established, but it is not mentioned as present in Italy in Noyes (2002).

1.8.7 *Fopius arisanus* (Sonan) (Hym., Braconidae)

Syn.: *Opius oophilus* Fullaway

Fopius arisanus is an endoparasitoid from the Malaysian area (Clausen 1978). It is known from 11 host species from four Tephritidae genera (mainly *Bactrocera*; Yu *et al.* 2005). Besides Europe, it was introduced into Israel, Peru, Argentina and Central America against *C. capitata* (BIOCAT 2005). *Fopius arisanus* was also released against other Tephritidae pests in Australia, Fiji, Hawaii, Madagascar, Marianas (Guam, Saipan), Mauritius, Mexico, New Zealand, the Philippines and the USA (BIOCAT 2005).

In 1959, *F. arisanus* obtained from Hawaii was released in Italy, but the species failed to establish (Clausen 1978). It was further released in 1966 on Rhodes, Greece, but again failed to establish (Greathead 1976).

1.8.8 *Halticoptera daci* Silvestri (Hym., Pteromalidae)

Halticoptera daci is an ectoparasitoid native to Eritrea (Greathead 1971). It was collected from the soil underneath olive trees in South Africa, implying that it might be emerging from olive fruit fly pupae (Neuenschwander 1982). No other hosts are reported for this species (CABI 2007). However, given the low number of *H. daci* collected in olive fruit fly surveys, Daane and Johnson (2010) argue that this parasitoid is likely to be a polyphagous species that opportunistically attacks olive fruit fly.

In 1914, *H. daci* was collected in Eritrea, and in 1914/15, 47 individuals were released in Italy (Clausen 1978). The species is reported as established in Italy (CABI 2007), but its impact on *B. oleae* is unknown.

1.8.9 *Mesopolobus modestus* Silvestri (Hym., Pteromalidae)

Syn.: *Amblymerus modestus* Silvestri
Mesopolobus modestus is an ectoparasitoid native to Eritrea (Greathead 1971). No other hosts are reported for this species (CABI 2007).

In 1914, *M. modestus* was collected in Eritrea, and in 1914/15, 20 individuals were released in Italy (Clausen 1978). Apparently, the species has established in Italy (CABI 2007), but it appears to be unsuccessful as a biological control agent because of the thickness of Italian olive skins (see also Section 1.8.3; Greathead 1971).

1.8.10 *Closterocerus formosus* Westwood (Hym., Eulophidae)

Syn.: *Neochrysocharis formosa* (Westwood); *Achrysocharis formosa* Westwood; *Achrysocharella formosa* Westwood
Closterocerus formosus is a solitary larval endoparasitoid native to the Afrotropical region (from Ethiopia to South Africa; WaspWeb 2004–2010). Known hosts include numerous species, mainly leafminers, in the order Diptera, Coleoptera, Hemiptera, Hymenoptera and Lepidoptera (Noyes 2002). In Europe, *C. formosus* is reported from Austria, Croatia, Czech Republic, Denmark, Finland, France, Germany, Greece, Hungary, Italy, Montenegro, the Netherlands, Poland, Serbia, the Slovak Republic, Spain, Sweden, Switzerland, the UK and former Yugoslavia (Noyes 2002; CABI 2007).

In 1914, *C. formosus* was collected in Eritrea, and in 1914/15, 400 individuals were released in Italy (Clausen 1978). The species has established but appears to be unsuccessful as a biological control agent, probably due to its short ovipositor that makes it difficult to penetrate the thick skin of Italian olives (Greathead 1971).

1.8.11 *Diachasmimorpha tryoni* (Cameron) (Hym., Braconidae)

Syn.: *Opius tryoni* Cameron; *Parasteres tryoni* (Cameron)
Diachasmimorpha tryoni is native to Australia (Carmichael *et al.* 2005). It is recorded from at least 18 Tephritidae species from six genera (mainly *Anastrepha*, *Bactrocera*

and *Ceratitis* species; Yu *et al.* 2005; Wharton and Yoder 2016). Besides Europe, *D. tryoni* has also been introduced against Tephritidae pest species into Cape Verde, Central America, Cook Islands, Egypt, French Polynesia, Guatemala, Israel, Mexico, Puerto Rico and the USA (California, Florida, Hawaii) (BIOCAT 2005; Wharton and Yoder 2016). *Diachasmimorpha tryoni* is one of the most important species used in tephritid biocontrol programmes to date. For instance, the species continues to play an important role in suppressing *C. capitata* in Hawaii (Wharton and Yoder 2016). Methods for mass rearing this species for augmentative biological control have been developed (Wharton and Yoder 2016). However, *D. tryoni* also attacks the gall-forming tephritid fly *Eutreta xanthochaeta* Aldrich that was introduced into Hawaii to control the invasive shrub *Lantana camara* L. (Duan *et al.* 2000).

In 1966, *D. tryoni* collected in Hawaii was released in Greece (Rhodes; H.G. Stavraki, personal comments, in Greathead 1976). A direct release of 1500 individuals of *Opius* spp. was made, but this number also comprised *F. arisanus* (Sonan) and *B. longicaudatus*. The species is reported as established in Greece (CABI 2007). Moreover, *D. tryoni* is also recorded as introduced (van Achterberg 1999) and established in Spain (Yu *et al.* 2005). No information on its success as a biocontrol agent in Greece or Spain has been found.

1.8.12 *Phaedrotoma trimaculata* (Spinola) (Hym., Braconidae)

Syn.: *Opius trimaculatus* Spinola
Phaedrotoma trimaculata is native to the Neotropical region (Yu *et al.* 2005). It is also recorded from *Anastrepha fraterculа* Wiedemann, *C. capitata* and *Drosophila flavopilosa* Frey (Yu *et al.* 2005; Wharton and Yoder 2016). Besides Europe, it has also been introduced against *C. capitata* in South Africa (Yu *et al.* 2005).

In 1937, *P. trimaculata* from Brazil was brought to Italy, where a total of 200 laboratory-reared individuals were released (Silvestri 1938). Although originally reported as not established (Greathead 1976), the species has recently been recorded from Italy (CABI 2007). No information is available on its success as a biocontrol agent.

1.8.13 *Psyttalia concolor* Szépligeti (Hym., Braconidae)

Syn.: *Opius concolor* Szépligeti
Psyttalia concolor is a larval-pupal parasitoid native to many countries in Africa (WaspWeb 2004–2010). The discovery of *P. concolor* as a natural parasitoid of *B. oleae* was first made by P. Marchal in Tunisia in 1910 (Greathead 1976). The species is recorded from at least 12 Tephritidae species from eight genera (Yu *et al.* 2005). Besides Europe, this species has also been introduced against other tephritid pests (*Anastrepha striata* Schiner, *Anastrepha suspensa* (Loew), *Bactrocera tryoni* (Froggatt), *C. capitata*, *Ceratitis malagassa* Munro and *Dacus frontalis* Becker)

into Australia, Bermuda, Bolivia, Cape Verde Islands, Costa Rica, Ecuador, Israel, Jordan, Lebanon, Madagascar, New Caledonia, Palestinian Territory, Pakistan, Puerto Rico, La Réunion, Trinidad and USA (Florida and Hawaii) (BIOCAT 2005; Wharton and Yoder 2016). In Europe, *P. concolor* is reported as present in France, Greece, Italy and Spain (Fischer and Koponen 1999; CABI 2007).

In 1914, 112 individuals of *P. concolor* collected in Tunisia were released by F. Silvestri in Calabria, Italy. Between 1917 and 1927, Silvestri made additional releases in different parts of southern Italy, including Sicily (Silvestri 1938). Once this species was found to have established in Sicily, Silvestri started collecting material in Sicily and releasing it on the mainland (Silvestri 1938). In the late 1950s, *P. concolor* was also introduced into northern Italy, including Tuscany and the island of Elba (Monastero 1960). In 1960, a programme of mass rearing of *P. concolor* was initated to assess experimentally the effects of annual mass releases of this parasitoid on *B. oleae* (Monastero and Genduso 1962). For example, in a large-scale experiment on the Gargano Peninsula, over 11 million *P. concolor* were released in the first year (Genduso and Ragusa 1968).

First attempts to etablish *P. concolor* in France were made between 1919 and 1923 using insects from Tunisia, but adults reared from infested fruits either failed to breed or were released but no recoveries were made (Greathead 1976). In 1959, individuals reared from material collected in Morocco were released, and this time *P. concolor* established successfully (Feron *et al.* 1961). In 1961, the importance of hyperparasitism by native chalcidoid parasites was discovered, strongly reducing the effect of *P. concolor* (Delanoue 1970).

In 1923, *P. concolor* was also released in Spain using specimens collected in Tripolitania (Libya). The species bred in the field, but did not survive the winter (Aguilò 1924). Additional releases were made in 1968/69 and in 1985 with specimens obtained from France (Jiménez *et al.* 1969).

The first releases of *P. concolor* in Greece were made in 1933 on Crete using individuals from Italy. Etablishment of this species on Crete was confirmed in 1963 (Pelekassis 1974). Additional releases were made in 1965 (25,000 individuals originating from France) and in 1966/67 (400,000 individuals) on the island of Khalki, from where it spread naturally to Rhodes, a distance of over 9 km (Pelekassis 1974). Between 1965 and 1972, the parasitoid was released at seven additional locations, both on islands and on the mainland (Pelekassis 1974).

In 1968, 12,000 *P. concolor* were released in former Yugoslavia near Makarska in central Dalmatia (Brnetic 1968).

Overall, *P. concolor* has proven to be quite successful in controlling *B. oleae*. For instance, 50–60% parasitism was recorded at the release site in former Yugoslavia 1 month after release (Brnetic 1968), up to 54% on Elba (Fenili and Pegazzano 1962) and 83% in an experimental study (Monastero and Genduso 1962). However, high levels of parasitism were obtained mainly when regular releases were made; populations usually dropped to low levels when releases were stopped (Greathead 1976).

1.8.14 *Psyttalia dacicida* (Silvestri) (Hym., Braconidae)

Syn.: *Opius dacicida* Silvestri

Psyttalia dacicida is an endoparasitoid native to southern and south-eastern Africa and is listed among the braconid parasitoids providing the highest levels of olive fruit fly suppression in its native range (Daane and Johnson 2010). *Bactrocera oleae* is the only known host (Wharton and Yoder 2016).

In 1914, individuals collected in Eritrea were released in Italy (Silvestri 1938), where the species established but remained unsuccessful as a biological control agent (Yu *et al.* 2005). According to Greathead (1971), this species' ovipositor is also poorly adapted to parasitize its host through the thick skin of Italian olives.

1.8.15 *Triaspis daci* (Szépligeti) (Hym., Braconidae)

Triaspis daci is an endoparasitoid native to several African countries, including Democratic Republic of Congo, Ethiopia and South Africa (WaspWeb 2004–2010). The only recorded Tephritidae host is *B. oleae*; however, *T. daci* is also likely to be a parasitoid of Coleoptera species (Daane and Johnson 2010).

In 1914, individuals collected in Eritrea were released in Italy, but the species failed to establish (Silvestri 1938). To date, *T. daci* is recorded from Spain (Yu *et al.* 2005).

1.8.16 *Utetes africanus* (Szépligeti) (Hym., Braconidae)

Syn.: *Opius africanus* Szépligeti; *Opius tephritivorus* Wharton

Utetes africanus is an endoparasitoid native to many African countries (Eritrea, Ethiopia, Kenya, Senegal and South Africa; WaspWeb 2004–2010). It is also recorded from five other Tephritidae species, i.e. *Carpomya incomplete*, *Ceratitis capitata*, *Ceratitis rosa*, *Pterandrus rosa* and *Trirhithrum coffeae*, and from the agromyzid fly *Ophiomyia phaseoli* (Tryon) (Yu *et al.* 2005), but these records might need confirmation as they may apply to a different species of *Utetes* that was undescribed at the time the collections were made (Wharton and Yoder 2016). In South Africa, *U. africanus* has been most commonly reared from wild olives and it tends to be rare in cultivated olives (Daane and Johnson 2010).

In 1910, a small number of *U. africanus* collected in South Africa was released in Italy (Clausen 1978). In 1911 and 1914, collections were made again in Eritrea, and 440 individuals were released in Italy (Clausen 1978). The species has established (Yu *et al.* 2005), but it has been unsuccessful as a biological control agent, probably because its short ovipositor is not adapted to the thick skins of Italian olive (Greathead 1971).

Utetes africanus was further considered for shipping from South Africa to Spain (Greathead 1976), but no information was found on an actual introduction into Spain.

1.9 *Ceratitis capitata* (Wiedemann), Mediterranean Fruit Fly (Dip., Tephritidae)

This pest species is native to sub-Saharan Africa, from where it has spread to Europe, Egypt, the Middle East, the Malagasy subregion, Australia and the Americas. The transport of infested fruits is the major means of movement and dispersal of this species. *Ceratitis capitata* has been intercepted in most European countries; however, permanent establishment is recorded mainly from Mediterranean countries, and it is widespread in Cyprus, Greece, Italy, Malta, Portugal and Spain. Restricted distribution is also reported for Switzerland.

Ceratitis capitata is a highly polyphagous species. Its pattern of host use varies from region to region and appears to relate largely to what fruits are available. Major hosts in Europe include apple, *Citrus* species, figs and stone fruits (*Prunus*). Its host range also includes wild species in the invaded range; for instance, box thorn (*Lycium europaeum* L.) is an important overwintering host in North Africa.

Larvae of *C. capitata* develop in fruits. Direct damage to fruit crops is frequently high and may reach 100%. Moreover, adults may also transmit fruit-rotting fungi.

To protect crops from *C. capitata*, cover sprays of entire crops are sometimes used, but bait sprays (i.e. insecticide mixed with a protein bait) are both more economic and more environmentally acceptable. Mechanical control consists of wrapping fruit in paper or polythene sleeve. The sterile insect technique has been used against *C. capitata* in Costa Rica, Italy, Mexico, Nicaragua, Peru, Spain, Tunisia and the USA (California and Hawaii). Natural enemies recorded on *C. capitata* include 40 parasitoid species from 21 genera, six predator and four pathogen species.

1.9.1 *Aceratoneuromyia indica* (Silvestri) (Hym., Eulophidae)

Syn: *Syntomosphyrum indicum* Silvestri

Aceratoneuromyia indica is a gregarious endoparasitoid from the Indo-Pacific region (Ovruski *et al.* 2006) and was first collected in India (Clausen 1978). Known hosts include members of the Tephritidae genera *Anastrepha*, *Bactrocera*, *Dacus* and *Ceratitis* (Wharton and Yoder 2016). Besides Europe, this species has been introduced into Australia, the Caribbean islands, Mexico, South Africa and Central and South America as a biological control agent of various fruit fly species of economic importance (Ovruski *et al.* 2006). In Europe, *A. indica* is reported from Italy and the UK (Noyes 2002; Rasplus *et al.* 2010).

In 1909, three releases of *A. indica* were made against *C. capitata* in southern Italy (Rosarno, Calabria; Silvestri 1938). In total, 10,000 individuals originating from Western Australia were released, but according to Greathead (1976), the species failed to establish.

1.9.2 ***Belonuchus rufipennis*** (Fabricius) (Col., Staphylinidae)

The native range of the predatory beetle *B. rufipennis* ranges from north-eastern USA to Argentina (Frank and Barreray 2010). Known hosts include Tephritidae and Drosophilidae (Diptera; Greathead 1976), but also the sap-feeding beetle *C. humeralis* (Col., Nitidulidae), against which it was released as a biocontrol agent in Hawaii (BIOCAT 2005). In Europe, the species was also released against *B. oleae* in Italy (Section 1.8.1; Greathead 1976).

Between 1939 and 1941, *B. rufipennis* originating from Brazil was released at 15 locations in mainland Italy and in Sicily (Silvestri 1945). The number of individuals released at each occasion ranged from 48 to 1540 (Silvestri 1945). The species established successfully in Sicily, but in 1959 no significant impact on the pest species was detected (Russo 1959).

1.9.3 ***Dirhinus giffardii*** Silvestri (Hym., Chalcididae)

Dirhinus giffardii is a larval pupal endoparasitoid native to Africa (Wang and Messing 2004). Known hosts include at least 19 Diptera species, mainly Tephritidae (Waterhouse 1993). Worldwide, the species was introduced into Australia, Costa Rica, Fiji, Israel, Micronesia, Pakistan, Puerto Rico and the USA (Hawaii) for biological control of various Diptera species (Noyes 2002).

In 1912, releases of *D. giffardii* against *C. capitata* were made in Italy (both mainland and Sicily; Silvestri 1914), where the species established (Noyes 2002). In 1962, it was also released in Greece, but failed to establish (Greathead 1976).

1.9.4 ***Coptera silvestrii*** (Kieffer) (Hym., Diapriidae)

Syn.: *Psilus silvestrii* (Kieffer)

Coptera silvestrii is native to western Africa (Ovruski *et al.* 2006). Known hosts also include *Ceratitis punctata* (CABI 2007). Besides Europe, the species was also released in Hawaii against *C. capitata* (BIOCAT 2005).

In 1912, *C. silvestrii* was released in southern Italy and Sicily (Silvestri 1914). No information on the establishment and impact of the species has been found.

1.9.5 ***Tetrastichus giffardianus*** Silvestri (Hym., Eulophidae)

Tetrastichus giffardianus is a gregarious endoparasitoid from Africa (Waterhouse 1993). Known hosts include species in 11 Tephritidae genera (Noyes 2002). Besides Europe, it has been introduced in Argentina, Brazil, Central America, Egypt, Hawaii and Peru against *C. capitata* and established in all countries except Argentina and Peru (BIOCAT 2005). The species was also released against other Tephritidae pests (BIOCAT 2005). A recent study on non-target effects on an endemic tephritid *Trupanea dubautiae* (Bryan) in Hawaii revealed that *T. giffardianus* was able to parasitize late instars of *T. dubautiae* and to develop successfully to adults in the laboratory (Duan and Messing 1998). However, no *T. dubautiae* were attacked by *T. giffardianus* when presented in their respective host, and no *T. giffardianus* were recovered from *C. capitata* or *T. dubautiae* in extensive field surveys, indicating that the parasitoid was unlikely to affect either species under natural conditions.

In 1959, *T. giffardianus* was introduced into Spain, from where a good level of impact was reported (Greathead 1976). However, Noyes (2002) does not list this species as being present in Spain.

1.9.6 *Elachertus giffardi* Timberlake (Hym., Eulophidae)

Elachertus giffardi is a gregarious endoparasitoid from Africa (Noyes 2002). Known hosts include nine Tephritidae species (Noyes 2002).

According to Russo (1959), this species was introduced from East Africa into Italy. *Elachertus giffardi* is considered established in this country according to CABI (2007), but not according to Noyes (2002). No information on its impact was found.

1.10 *Choristoneura murinana* (Hübner), European Fir Budworm (Lep., Tortricidae)

Choristoneura murinana is considered to be native to Europe and is recorded as present in former Czechoslovakia, France, Germany, Italy, Poland, Russia and Switzerland (Sarýkaya and Avcý 2006). Damage caused by *C. murinana* was first recorded in 1805 in the Thuringian Forest in Germany; outbreaks of the insect were subsequently recorded in different regions of central and eastern Europe (Sarýkaya and Avcý 2006).

Choristoneura murinana is recorded from different *Abies* species and from *Cedrus atlantica* Endl. and *Picea abies* L. (Sarýkaya and Avcý 2006). This species feeds on the needles, buds and shoots. After heavy defoliation, the tree weakens and is attacked by bark beetles, which may cause the death of the tree (Sarýkaya and Avcý 2006).

Chemical insecticides (such as malathion, carbaryl and acephate), as well as microbial insecticides (such as the bacterium *Bacillus thuringiensis* Berliner), can be used to reduce spruce budworms (*Choristoneura* spp.) in general.

Parasitoids recorded on *C. murinana* include eight Ichneumonidae, one Encrytridae, one Trichogrammatidae and two Braconidae species (all Hymenoptera), as well as four Tachinidae species (Diptera). In addition, *Formica polyctena* Förster (Formicidae, Hymenoptera) is recorded as a predator of the European fir budworm.

1.10.1 *Apanteles fumiferanae* Viereck (Hym., Braconidae)

Apanteles fumiferanae is an endoparasitoid native to North America (Stein 1981). Host species include numerous Lepidoptera species from 20 genera (including Hesperidae, Pyralidae/Crambidae and Tortricidae; Yu *et al.* 2005).

In 1977/78, *A. fumiferanae* was released in Poland (Pruszynski 1989). Apparently, the species established (Yu *et al.* 2005), but no information on its impact was found.

1.10.2 *Ceromasia auricaudata* Townsend (Dip., Tachinidae)

Ceromasia auricaudata is a larval endoparasitoid native to North America (O'Hara and Mahony 2009a). Host species include two other *Choristoneura* species, *Choristoneura fumiferana* (Clemens) and *Choristoneura occidentalis* (Walsingham), as well as other Lepidoptera species in the families Geometridae, Noctuidae, Pieridae and Pyralidae (CABI 2007; O'Hara and Mahony 2009a). Within North America, this species was moved from Canada to north-eastern USA and released as a biological control agent against *C. fumiferana* (BIOCAT 2005).

In 1977/78, *C. auricaudata* was released in Poland (Pruszynski 1989), where it apparently established (CABI 2007). No information on its success as a biocontrol agent was found.

1.10.3 *Glypta fumiferanae* (Viereck) (Hym., Ichneumonidae)

Glypta fumiferanae is an endoparasitoid native to North America (Stein 1981). Known hosts include 13 species from six genera (mainly Tortricidae, but also Pyralidae; Yu *et al.* 2005).

In 1977/78, *G. fumiferanae* was released in Poland (Pruszynski 1989), where it apparently established (CABI 2007); however, no confirmation of its establishment has been published since.

1.10.4 *Casinaria* sp. (Hym., Ichneumonidae)

Syn.: *Horogenes* sp.

Species in the genus *Casinaria* have been used worldwide as biocontrol agents against Lepidoptera (Tortricidae, Coleophoridae and Pyralidae; BIOCAT 2005).

In 1977/78, an unspecified *Casinaria* species from North America was released in Poland. It is not known whether or not it established successfully (Pruszynski 1989).

1.10.5 *Smidtia fumiferanae* Tothill (Dip., Tachinidae)

Syn.: *Omotoma fumiferanae* (Tothill)
Smidtia fumiferanae is a larval-pupal endoparasitoid native to North America (O'Hara and Mahony 2009c). In addition to at least four *Choristoneura* species, known hosts also include two species of Geometridae and several species of Noctuidae (O'Hara 2005). It is the most common Tachinidae parasitoid of *C. fumiferana* (Clem.) in British Columbia (O'Hara 2005).

In 1977/78, *S. fumiferanae* was released in Poland (Pruszynski 1989), where it apparently established (CABI 2007); no information on its success as a biocontrol agent was found.

1.11 *Chrysomphalus dictyospermi* (Morgan), Dictyospermum Scale (Hem., Diaspididae)

Chrysomphalus dictyospermi was probably native to southern China, but was spread a long time ago, presumably on *Citrus* planting material and fruit, so there is no history of its spread traceable in the literature. Today, it is widespread in tropical and subtropical regions and occurs under glass in temperate areas. In Europe, it was first recorded in Florence, Italy, in 1895 (Silvestri 1926), and then in southern France in 1899. To date, *C. dictyospermi* is recorded as present in Albania, Bosnia and Herzegovina, Bulgaria, Croatia, Estonia, France, Germany, Greece, Italy, Latvia, Lithuania, Former Yugoslav Republic of Macedonia, Malta, Montenegro, Poland, Portugal, Romania, Serbia and Spain.

Chrysomphalus dictyospermi is a highly polyphagous species and has been recorded from hosts belonging to 73 plant families. Favoured hosts are *Citrus* and other trees such as olives and palms. It prefers feeding on leaves but also attacks fruits and stems. *Chrysomphalus dictyospermi* is considered among the most serious pests of *Citrus*. For instance, it has been reported as responsible for the rejection of 22% of fruits in sorting and packing houses in Spain.

Chrysomphalus dictyospermi can be treated successfully with white mineral oils. Biological control of this species is also very effective. For instance, control of *C. dictyospermi* by natural enemies in Cuba is so effective that the species is only found in orchards that are treated with insecticides which reduce the natural enemy populations.

Overall, 13 parasitoid species from four genera (mainly *Aphytis* (Aphelinidae, Hymenoptera)) and three predatory species are recorded attacking *C. dictyospermi.*

1.11.1 *Aphytis coheni* DeBach (Hym., Aphelinidae)

Aphytis coheni is an ectoparasitoid attacking nymphs and adults. This species was originally described from Israel (CABI 2007), but it might be native to south-eastern Asia (Avidov *et al.* 1970). *Aphytis coheni* is recorded to attack various other scale insects (i.e. *A. aurantii*, *Aonidiella citrina* (Coquillett), *A. nerii* and *Hemiberlesia lataniae* (Signoret); Noyes 2002). Besides Europe, the species was also introduced into Australia, South Africa and the USA against another Diaspididae species, *A. aurantii* (BIOCAT 2005). In Europe, it was also introduced into Cyprus against *L. beckii* (Section 1.27.1).

In 1962, 7780 *A. coheni* collected in California were released in Greece (DeBach and Argyriou 1967). The species is reported as established according to CABI (2007), but not according to Noyes (2002). No information on its success as a biocontrol agent was found.

1.11.2 *Aphytis lingnanensis* Compere (Hym., Aphelinidae)

Aphytis lingnanensis is a gregarious ectoparasitoid native to China, India and Pakistan (CABI 2007). Known hosts include numerous Diaspididae species (Noyes 2002). Besides Europe, the species was also introduced into Argentina, Australia, Chile, Israel, Mexico, Morocco, South Africa and the USA against *A. aurantii* (BIOCAT 2005). It was further used as a biological control agent against nine other Diaspididae *Citrus* pests and introduced into Bermuda, Costa Rica, Cuba, Iran, Japan, Peru, Solomon Islands and the former USSR (BIOCAT 2005; CABI 2007). In most areas, *A. lingnanensis* has had a beneficial impact, alone or along with other introduced *Aphytis* spp. (CABI 2007). In Europe, the species is recorded from Albania, Cyprus, Greece, Italy and Spain (Noyes 2002; CABI 2007; Rasplus *et al.* 2010).

In 1962, 5650 individuals originating from the USA were released in Greece against both *C. dictyospermi* and *A. aurantii*, and an additional 14,000 individuals were released in Crete in 1963 (see also Section 1.4.2). The species failed to establish in both areas (Greathead 1976). In 1964, *A. lingnanensis* was also released in Sicily, Italy (Greathead 1976; Clausen 1978). Information about its establishment is contradictory: according to BIOCAT 2005, the species failed to establish, but it has been recorded in CABI (2007) as present in Italy.

1.11.3 *Aphytis melinus* DeBach, golden chalcid (Hym., Aphelinidae)

Aphytis melinus is a gregarious ectoparasitoid native to India and Pakistan (CABI 2007). Known hosts include at least 12 other Diaspididae species including many economically important ones (Noyes 2002), but also *A. nerii* (Aphidae, Hemiptera) and *S. citri* (Thripidae, Thysanoptera; CABI 2007). Besides Europe, *A. melinus* was also introduced into Argentina, Australia, Chile, China, Israel, Morocco, South Africa and the USA against *A. aurantii* (BIOCAT 2005). It was further used as a biological control agent against five other Diaspididae *Citrus* pests in Iran, Peru and the former USSR (BIOCAT 2005). In Europe, the species was also introduced against other scale insects in France, Italy, Greece and Spain (Sections 1.4.3, 1.7.1). To date, *A. melinus* is recorded in Europe from Albania, Belgium, Cyprus, Czech

Republic, Denmark, France, Germany, Greece, Italy, the Netherlands, Portugal, Spain and former Yugoslavia (Noyes 2002; CABI 2007; Rasplus *et al.* 2010).

In 1962, *A. melinus* was released together with other *Aphytis* species in Greece, against both *C. dictyospermi* and *A. aurantii* (Greathead 1976; see also Section 1.4.3). *Aphytis melinus* established in spite of the severe winter of 1962/63, and started to spread from 1963 at a rate of 75–100 km/year. Substantial to complete control of scale pests was achieved at the release sites (Greathead 1976; Greathead and Greathead 1992). In most areas where *A. melinus* was released, it largely displaced another introduced parasitoid against scale insects, *A. chrysomphali* (Greathead 1976). *Aphytis melinus* also established on *A. nerii*, a minor pest on olives and citrus (Section 1.7).

In 1964, *A. melinus* originating from California was released in Italy (Inserra 1966). The species established and rates of parasitism up to 37% were recorded. In the absence of Argentine ants (*Linepithema humile* (Mayr)), complete control of the pest is achieved (Inserra 1970). In 1966, *A. melinus* was introduced into France, where it also established (Malausa *et al.* 2008). In 1969, the species was released in Spain, from where substantial control of pest numbers was reported (Cavalloro and di Martino 1986; Pina *et al.* 2003; Pina and Verdú 2007). Virtually complete control was also reported from Corsica, where the species was released in 1972 (Greathead 1976).

1.11.4 *Comperiella bifasciata* Howard (Hym., Encyrtidae)

Comperiella bifasciata is a solitary, endoparasitoid native to eastern Asia (China) (CABI 2007). Known hosts include at least 21 Diaspididae species (Noyes 2002). *Comperiella bifasciata* was further used as a biological control agent against six other Diaspididae pests and was introduced into Argentina, Australia, Bermuda, Chile, Egypt, Israel, Mexico, New Zealand, Philippines, Swaziland, South Africa and the USA (BIOCAT 2005). In Europe, the species was introduced into Cyprus, France and Greece against other scale insects (Diaspididae and Coccidae; Sections 1.4.4, 1.56.6). To date, it has been recorded as present in Belgium, Cyprus, the Czech Republic, France, Greece, Hungary, Italy, Moldova, the Netherlands, Romania, Russia, Spain, Ukraine and former Yugoslavia (Noyes 2002; Rasplus *et al.* 2010).

In 1969, *C. bifasciata* originating from the USA was released in Greece (Greathead 1976). Since it failed to establish, a second release of individuals from France was made in 1972. The species established and is recorded as present in Greece (CABI 2007), but apparently has no major impact as a biological control agent, because complete control of *C. dictyospermi* has been achieved with *A. melinus* (see above).

1.11.5 *Encarsia lounsburyi* (Berlese and Paoli) (Hym., Aphelinidae)

Syn.: *Aspidiotiphagus lounsburyi* (Berlese and Paoli); *Prospaltella lounsburyi* Berlese and Paoli

Encarsia lounsburyi was first described from Madeira (Polaszek *et al.* 1999). Known hosts include 13 genera in the family Diaspididae (Heraty *et al.* 2007). Besides Europe, *E. lounsburyi* was introduced into Canada and the USA as a biological control agent against three other Diaspididae pests (BIOCAT 2005). In Europe, the species is recorded from Albania, Cyprus, France, Greece, Italy, the Netherlands, Portugal, Spain and Switzerland (Noyes 2002; Rasplus *et al.* 2010).

In 1921, *E. lounsburyi* originating from Madeira was released in Italy (Silvestri 1926). At first, the species was reported to be unable to control *C. dictyospermi*, but in 1959, its impact on the pest was considered 'of great importance' (Domenichini 1959). In 1923, individuals originating from Italy were released in France (Greathead 1976). Today, *E. lounsburyi* is recorded as present in France (CABI 2007), but no information on its success as biocontrol agent has been found.

1.11.6 *Encarsia perniciosi* (Tower) (Hym., Aphelinidae)

Syn.: *Prospaltella perniciosi* Tower

Encarsia perniciosi is an endoparasitoid native to North America (Polaszek *et al.* 1999). Known hosts include seven genera in the family Diaspididae (Heraty *et al.* 2007). Worldwide, the species was also used as a biological control agent against two other Diaspididae pests, *A. aurantii* and *Pseudaulacaspis pentagona* (Targioni-Tozzetti), and was introduced in Australia, Bermuda, Chile, India, Israel, New Zealand, Pakistan, South Africa, the USA and the former USSR (BIOCAT 2005). In Europe, *E. perniciosi* was also introduced in Austria, former Czechoslovakia, France, Germany, Greece, Italy, Yugoslavia, Spain and Switzerland to control *Diaspidiotus perniciosus* (Section 1.18.2). It is recorded as present in Albania, Austria, Bulgaria, former Czechoslovakia, Denmark, France, Germany, Greece, Hungary, Italy, Portugal, Romania, Spain, Switzerland, the former USSR and former Yugoslavia (Serbia and Montenegro) (Noyes 2002; Rasplus *et al.* 2010).

In 1969 and 1972, *E. perniciosi* originating from USA was released in Greece and France, respectively (Greathead 1976). The species is recorded as present in both countries (CABI 2007), but no information on its impact on *C. dictyospermi* has been found.

1.12 *Cydia molesta* (Busck), Oriental Fruit Moth (Lep., Tortricidae)

Syn.: *Grapholita molesta* (Busck); *Laspeyresia molesta* Busck

Cydia molesta is native to north-western China, from where it spread to Australia, central Europe, the east coast of the USA and Brazil at the beginning of the 20th century. Today, it has spread to practically all the major stonefruit-growing areas of the world and is also recorded in temperate South America and New Zealand. *Cydia molesta* has been recorded recently for the first time in South Africa (Blomefield and Geertsema 1990). International movement is likely to occur on fruit or plants for planting. *Cydia molesta* was first reported from Europe in Italy (Greathead 1976). Today, the species is recorded from Austria, Bulgaria, Croatia, Czech Republic, France, Germany, Greece, Hungary, Italy, Latvia, Lithuania, Malta, Moldova, Montenegro, Portugal, Romania, Serbia, Slovak Republic, Spain, Switzerland and the UK.

The principal economic hosts are fruit trees of the genera *Prunus*, *Malus* and *Pyrus*. *Cydia molesta* also occurs on other fruit trees and ornamental trees in the Rosaceae (*Cotoneaster*, *Crataegus*, *Cydonia*), and on wild species. *Cydia molesta* is a serious pest of economic importance in commercial orchards of peaches and apricots; occasionally, it can also attack other commercial fruits, feeding on the fruits, leaves and stems of its hosts. In years with severe attack, young trees can suffer distortion of growing shoots and stems. Attack on fruits considerably reduces their quality and market value. *Cydia molesta* damage also encourages brown-rot infection (*Monilinia* spp.).

Orchards can be protected against *C. molesta* by means of chemical control combined with cultural methods (removal of pruned material). Furthermore, mating disruption has also been used successfully in some commercial peach orchards (Audemard *et al.* 1989).

Natural enemies recorded on *C. molesta* include 34 parasitoid species from 26 genera and seven pathogen species (CABI 2007).

1.12.1 *Glypta rufiscutellaris* Cresson (Hym., Ichneumonidae)

Syn.: *Glypta caulicola* Cushman

Glypta rufiscutellaris is a larval endoparasitoid native to North America (Weaver 1949). It is recorded from seven Lepidoptera species, including Tortricidae and Gelechidae (CABI 2007). Besides Europe, the species was also introduced in Argentina, Australia, Japan and Uruguay against *C. molesta* and *Cydia nigricana* (F.), as well as into Hawaii against *Phthorimaea operculella* (Zeller) and *Cryptophlebia illepida* (Butler) (Gelechiidae, Lepidoptera; BIOCAT 2005; Yu *et al.* 2005). *Glypta rufiscutellaris* is also recorded from countries of the former USSR (CABI 2007).

In 1934, *G. rufiscutellaris* was released in Italy (Greathead 1976). The species has apparently established (Yu *et al.* 2005), but no information on its impact on *C. molesta* has been found.

1.12.2 *Hymenochaonia delicata* (Cresson) (Hym., Braconidae)

Syn.: *Macrocentrus delicatus* Cresson

Hymenochaonia delicata is a solitary endoparasitoid native to North America (Yu *et al.* 2005). It is recorded from numerous Lepidoptera species from 17 genera, including Gelechiidae, Noctuidae, Pyralidae and Tortricidae (Yu *et al.* 2005). Besides Europe, this species has also been introduced into Australia against *C. molesta* (BIOCAT 2005).

In 1934, *H. delicata* originating from the USA was released in Italy (Greathead 1976), where it established successfully (Yu *et al.* 2005). The species has also been reported from France (Rasplus *et al.* 2010), but no information on its impact on *C. molesta* has been found.

1.12.3 *Macrocentrus ancylivorus* Rohwer (Hym., Braconidae)

Macrocentrus ancylivorus is a poly-embryonic parasitoid native to the north-eastern USA (LeRoux 1971), which was originally recorded from the North American strawberry leaf roller, *Ancylis comptana* (Froel.). Its host range includes numerous Lepidoptera species from 21 genera (including Gelechiidae, Noctuidae and Pyralidae; Yu *et al.* 2005). Besides Europe, *M. ancylivorus* has also been introduced into Argentina, Australia, Brazil, Canada, Chile, Columbia, Japan, Uruguay, parts of the USA and the former USSR against *C. molesta* and *C. nigricana* (BIOCAT 2005; Yu *et al.* 2005), as well as into Puerto Rico against *Etiella zinckenella* (Treitschke) (Pyralidae; BIOCAT 2005). In Europe, the species is recorded from Italy and Switzerland (Yu *et al.* 2005).

In 1931, three releases of *M. ancylivorus* were made in southern France, releasing 64, 200 and 785 individuals at a time (Greathead 1976). In 1953, an additional 30,000 individuals were released (Vayssière 1953). These individuals were obtained from a rearing culture based on *M. ancylivorus* pupae shipped from New Jersey (USA) and cultured on potato tuber moth, *P. operculella* (Zeller). Initial establishment was recorded at one site in France (Greathead 1976), but the species is thought to have subsequently died out.

In 1933, 600 cocoons of *M. ancylivorus* from USA were released in Italy (Bologna). The species was again released in 1951 at Salerno (Domenichini 1959). While Domenichini (1959) reported that these introductions failed, Yu *et al.* (2005) considered *M. ancylivorus* as established in Italy. No information on its impact on *C. molesta* was found.

1.13 *Cydia pomonella* L., Codling Moth (Lep., Tortricidae)

Cydia pomonella is native to Europe and has become a major pest of apples, pears and walnuts in most countries where cultivars of these species are grown (Barnes 1991). The females lay eggs singly on fruit or leaves in summer. Larvae hatch after 3–7 days and start burrowing into the fruits or nuts.

Until the 1990s, codling moth had been managed effectively through the intensive use of organophosphate insecticides. However, increasing regulatory pressure and evidence of cross-resistance to a number of insecticides highlighted the need to develop sustainable management systems for codling moth that use more diverse control tactics. To date, pheromone-based mating disruption has become an important component of pest management programmes targeting codling moth. Moreover, mass releases of commercially produced *Trichogramma* wasps and the application of *C. pomonella* granulovirus (CpGV) as a bioinsecticide have been shown to reduce damage from codling moth significantly (Mills *et al.* 2000; Lacey *et al.* 2008).

Natural enemies recorded on *C. pomonella* include more than 100 parasitoid species and a large number of entomopathogens (Lacey and Unruh 2007).

1.13.1 *Trichogramma minutum* Riley, minute egg parasite (Hym., Trichogrammatidae)

Trichogramma minutum is an egg parasitoid native to North America (Stouthamer *et al.* 2000), which has been recorded from 56 Lepidoptera species from different families (CABI 2007). Besides Europe, *T. minutum* was also introduced as a biocontrol agent into Chile, Ecuador, Egypt, Fiji, Greece, India, Papua New Guinea, the Philiippines, New Zealand and Sri Lanka against several Lepidoptera pests (Gelechiidae, Pyralidae, Noctuidae, Oecophoridae Tortricidae, Zygaenidae) (BIOCAT 2005). In Europe, it was also introduced against *Panolis flammea* (Denis and Schiffermüller) (Section 1.37.1) and *Prays oleae* (Section 1.46.3) and has been recorded as present in the Czech Republic, France, Germany, Greece, Italy, Spain and the UK (CABI 2007; Rasplus *et al.* 2010).

Attempts to control codling moth with *T. minutum* were made in Spain in the1930s and 1940s (Urquijo 1951). In 1935 and 1936, shipments of *T. minutum* were received from the USA, and first releases were made in 1936 on apple trees. An additional shipment made in 1940 failed, but success was finally achieved by a shipment made in 1941 containing some 10,000 eggs of the grain moth *Sitotroga cerealella* (Olivier) (Lep., Gelechiidae) that were parasitized by *T. minutum.* Releases were made in 1942 at a rate of 5000–20,000 parasitized eggs. By 1946, infestations of *C. pomonella* were reduced to 16–17% on the trees on which *T. minutum* was released, to 18–25% on neighbouring trees and to 50% on more distant trees (Urquijo 1951).

1.14 *Cinara laportei* (Remaudière) (Hem., Aphididae)

Cinara laportei is native to North Africa (Greathead 1971; Laamari *et al.* 2010). This species lives on *C. atlantica* (Laamari *et al.* 2010).

In 1967, *C. laportei* was introduced into the Mediterranean region of France, where it caused major damage to cedar plantations (Laamari *et al.* 2010). In 1971, it was also found in Italy, in 1972 in Spain, in 1974 in England, in 1975 in Germany and in 1976 in the Netherlands (Greathead 1971).

1.14.1 *Pauesia cedrobii* Starý and Leclant (Hym., Braconidae)

According to Greathead and Greathead (1992), *Pauesia cedrobii* is native to Morocco. The species is only recorded from *C. laportei* (Yu *et al.* 2005).

In 1981, a few hundred *P. cedrobii* individuals originating from Morocco were released in France (Fabre and Rabasse 1987; Malausa *et al.* 2008). The species established and spread quickly to large parts of France (Fabre and Rabasse 1987). In 1991, an additional introduction of *P. cedrobii* collected in Algeria was made (Mouna and Fabre 2005). Overall, this biological control project has been considered successful (Mouna and Fabre 2005).

1.15 *Ctenarytaina eucalypti* (Maskell), Blue Gum Psyllid (Hem., Psyllidae)

Ctenarytaina eucalypti is native to Australia. The species has been introduced into most of the eucalyptus-growing areas of the world and is now recorded from Africa, Asia, Europe and North and South America. In Europe, *C. eucalypti* was introduced into the British Isles as early as the 1920s, and into France at about the same time. The species is moved easily on *Eucalyptus* grown in commercial nurseries. To date, it is recorded as present in France, Germany, Ireland, Italy, Portugal, Spain and the UK.

Ctenarytaina eucalypti is specific on *Eucalyptus* species; two species, *Eucalyptus globulus* Labillardière and *Eucalyptus pulverulenta* Sims, are recorded as its major hosts. *Ctenarytaina eucalypti* feeds on leaves and stems, which results in leaf discoloration, rolling of leaves and twig distortion. In heavy infestations, it causes severe twig dieback. Damage is probably most severe in the plantations where the trees are grown for floral trade. In addition, the nymphs produce a large amount of honeydew that is a substrate for sooty mould.

Prior to the introduction of *C. eucalypti*, *Eucalyptus* plantations required very little upkeep and few pest problems occurred. The introduction of *C. eucalypti* caused heavy damage and control costs were estimated in the USA to be as high as US$45/acre for ground spray and US$100/acre for treatment by helicopter.

Chemical control has been used but insecticidal sprays are costly and considered ineffective. Biological control has been carried out worldwide and has resulted in effective control in several places.

Natural enemies recorded on *C. eucalypti* include one parasitoid species (*Psyllaephagus pilosus* Noyes) and four predatory species.

1.15.1 *Psyllaephagus pilosus* Noyes (Hym., Encyrtidae)

Psyllaephagus pilosus is a larval or nymphal endoparasitoid native to Australia and New Zealand that seems to be specific to *C. eucalypti* (Hodkinson 1999; Noyes 2002). Besides Europe, it has also been introduced into the USA to control *C. eucalypti* (BIOCAT 2005). Apparently, no host-range testing was conducted before releasing the species in Europe. However, the chances of *P. pilosus* attacking native psyllid species on a north temperate host are considered remote because of: (i) the taxonomically and chemically isolated position of the host plant of the psyllid (i.e. *Eucalyptus*); and (ii) the major size mismatch between *P. pilosus* and most potential alternative psyllid hosts (Hodkinson 1999). Besides the countries mentioned below, the species is also present in Italy, to where it is thought to have spread naturally from France (CABI 2007).

In 1994, *P. pilosus* originating from the USA (California) was released in the UK (CABI 2007); 300 individuals were released directly in the field and 200 additional ones liberated into cages covering infested *Eucalyptus neglecta* Maiden plants (Hodkinson 1999).

In April 1997, 463 and 200 *P. pilosus* individuals originating from the USA were released in France (Malausa and Girardet 1997). In 1998, 200 adults and 2000 mummified psyllid larvae from France were introduced into Ireland (Hodkinson 1999; Chauzat *et al.* 2001, 2002). *Psyllaephagus pilosus* has proven highly effective in reducing *C. eucalypti* populations and in preventing mass outbreaks wherever it has been introduced (Hodkinson 1999). In many places, there is no longer any need for chemical control, resulting in significant cost benefits. Data available from the USA indicate benefit:cost ratios between 9:1 and 24:1 (Hodkinson 1999). Problems of winter survival by *P. pilosus* might, however, limit the use of this insect in more northern regions (Hodkinson 1999).

1.16 *Dendroctonus micans* (Kugelann), Great Spruce Bark Beetle (Col., Curculionidae)

Dendroctonus micans is believed to originate from the conifer forests of Asia. Aided by the increased trade of unprocessed logs, the species has spread steadily westward over the past 100 years. At present, it is found throughout Eurasia and has adapted to a wide range of forest conditions. *Dendroctonus micans* is recorded as present in Austria, Belgium, Bosnia and Herzegovina, Bulgaria, Croatia, the Czech Republic, Denmark, Estonia, Finland, France, Germany, Hungary, Italy, Latvia, Lithuania, Luxembourg, the Netherlands, Norway, Poland, Romania, Serbia and Montenegro, the Slovak Republic, Sweden, Switzerland, Ukraine and the UK.

Dendroctonus micans breeds primarily in spruce (*Picea* spp.), especially in *P. abies* (L.) H. Karst. (Norway spruce), *Picea sitchensis* (Bong.) Carr. (Sitka spruce) and *Picea orientalis* (L.) Link (Oriental spruce). It can, however, also attack *Abies* species, *Larix decidua* Mill. (common larch) and *Pseudotsuga menziesii* (Mirbel) Franco (Douglas-fir). *Dendroctonus micans* larvae and adults feed on the cambium layer and inner bark. This species normally occurs in mature spruce forests; within its natural range, it causes low levels of mortality and is not considered to be a major pest. Widespread outbreaks, however, are common, especially along the edge of the *D. micans* range. Sustained attack of individual trees can result in tree mortality; in some cases, more than 50% mortality has occurred. Outbreaks are often more common in forests that are stressed; for example, by drought, poor soil nutrition and logging damage.

Insecticide application to infested portions of the lower boles of trees has been undertaken, but with questionable success. Integrated pest management of this insect consists of the timely detection of infestations, the rapid removal and processing of infested trees, thinning of overstocked stands to reduce their susceptibility to attack and the release of the predator, *Rhizophagus grandis* Gyllenhal, into areas where the target pest has recently spread.

Natural enemies reported from *D. micans* include at least five parasitoid species (four Hymenoptera and one Diptera species), two pathogen species of different genera and 11 predatory species. Woodpeckers prey on the larvae and pupae.

1.16.1 *Rhizophagus grandis* Gyllenhal (Col., Monotomidae)

Rhizophagus grandis is a predatory beetle native to Siberia (Gilbert and Gregoire 2003). In Europe, it seems to be monophagous on *D. micans*. In the native range, it is also reported as a natural enemy of *Dendroctonus valens* LeConte and the Curculionidae *Pissodes strobi* (Peck) (CABI 2007).

Rhizophagus grandis has expanded its range simultaneously with *D. micans*; however, its dispersal ability is lower than that of its prey. It has been estimated that *R. grandis* spreads at 200 m/year, compared to 1–10 km/year for *D. micans* (Gilbert and Gregoire 2003). Outside of Europe, *R. grandis* has also been introduced into Georgia, where it has contributed substantially to the control of the pest, and into Turkey (BIOCAT 2005). In Europe, the species is recorded as present in Belgium, France, Italy, the UK and the former USSR (CABI 2007), but it is likely to be more widespread.

In 1983, 200 specimens were imported into the UK to establish a rearing (Fielding and Evans 1997). During the same year, *R. grandis* was released at three sites and established successfully in two (Fielding and Evans 1997). As few as 27 pairs released resulted in establishment (Fielding and Evans 1997). Fielding and Evans (1997) report on host-specificity tests prior to releases: 'extensive testing of the predator was made against many other species of bark beetles, all of which were unaffected by this introduced predator'; however, no source for this information is provided. A mass release programme was initiated, and between 1984 and 1995, 156,358 individuals were released in 2741 sites (for more details see Fielding and Evans 1997). Substantial control of pest numbers is reported, and 13 years after its discovery, *D. micans* is considered under control in the UK (Fielding and Evans 1997).

In 1978, a first release of 2000 individuals was made in France (Van Averbeke and Grégoire 1995); some impact on pest number has been reported (Van Averbeke and Grégoire 1995). Since then, hundreds of thousands of specimens have probably been released in France to follow the spread of the bark beetle.

1.17 *Dialeurodes citri* (Ashmead), Citrus Whitefly (Hem., Aleyrodidae)

Dialeurodes citri is believed to be of oriental origin and is probably native to India. It has been widely spread across the world through the transport of living plants or fresh foliage. Today, it is found in south-eastern Asia, the Middle East, the Mediterranean region, the USA, Central and South America between the northern and southern 45th parallel of latitude. Outside of this zone, infestations by *D. citri* have been detected only under glass and on imported *Citrus* material. In Europe, the species is recorded as present in Cyprus, France, Greece, Italy, Malta and Spain.

Dialeurodes citri is a polyphagous species, with a host range comprising about 80 plant species from more than 50 genera and 30 families. It is an important pest of *Citrus*, coffee, *Diospyros kaki* Thunb. and a number of ornamentals. Besides cultivated plants, it is also recorded on *Ligustrum* (privet) in the wild. *D. citri* feeds on fruits, inflorescence, leaves and stems. It is one of the most important pests of *Citrus* spp. Serious damage is caused through the extraction of large quantities of sap and the development of sooty mould on the abundant excreted honeydew. In the case of heavy infestations, the whole canopy of a tree may acquire an almost completely black appearance.

Dialeurodes citri can be treated successfully with oil or oil emulsions. However, chemical applications usually provide only a temporary suppression and should be considered only for heavy infestations. In addition, insect growth regulators such as buprofezin and pyriproxyfen have proven to be effective on some stages of the pest.

A large number of natural enemies are known to develop on *D. citri*; some of them are of relevance to biological control. Overall, ten parasitoids in three genera of Hymenoptera, five pathogens and four predatory species are recorded on *D. citri*.

1.17.1 *Encarsia formosa* Gahan, whitefly parasite (Hym., Aphelinidae)

Encarsia formosa, an endoparasitoid, was first described from a specimen collected in a greenhouse in Idaho, USA. Its native range is uncertain (Hoddle *et al.* 1998). Known hosts are Aleyrodidae, including *D. citri*, *Trialeurodes vaporariorum* (Westwood), *Aleuroglandulus malangae* Russell, *Aleurotrachelus trachoides* (Back), *Aleyrodes lonicerae* Walker, *Aleyrodes proletella* L., species in the *Bemisia tabaci* complex, and *Dialeurodes chittendeni* Laing (CABI 2007). Besides Europe, the species was also released in Egypt to control *B. tabaci*, and in Australia, Israel and the USA (Hawaii), it was released against *T. vaporariorum* (BIOCAT 2005). In Europe, *E. formosa* is recorded as present in Albania, Austria, Belgium, Bulgaria, former Czechoslovakia, Denmark, Estonia, Finland, France, Germany, Greece, Hungary, Ireland, Italy, Malta, Moldova, the Netherlands, Norway, Poland, Portugal, Romania, Serbia, Slovenia, Spain, Sweden, Switzerland, the UK and the former USSR (Noyes 2002; CABI 2007; Rasplus *et al.* 2010). Since this species was and still is used also for biological control in greenhouses it is not clear whether it has permanently established in the environment in all these countries.

Before 1950, *E. formosa* originating from Belgium and Switzerland were released against *D. citri* in France (Greathead 1976). Single specimens have subsequently been recovered (Greathead 1976) and the species is recorded as present in France (CABI 2007). No information on its success as a biocontrol agent in the field was found.

1.17.2 *Encarsia lahorensis* (Howard) (Hym., Aphelinidae)

Encarsia lahorensis, a primary endoparasitoid, is native to India (Barbagallo *et al.* 1992). Known hosts also include three other Aleyrodidae, i.e. *Aleurocanthus spiniferus*, *Dialeurodes citrifolii* and *Trialeurodes ricini* (Noyes 2002). Besides Europe, it was also released against *D. citri* in Israel, Turkey, USA and the former USSR (BIOCAT 2005). *Encarsia lahorensis* is recorded in Europe in France, Greece, Italy and the former USSR (CABI 2007).

Between 1973 and 1975, *E. lahorensis* was released in mainland Italy, from where very good results in pest control were recorded (Barbagallo *et al.* 1992). In 1980/81, it was also released in Sardinia, with some impact on pest numbers (Ortu and Prota 1986). In 1976, *E. lahorensis* from California was released in both mainland France and Corsica (Malausa *et al.* 2008). The species has established in mainland France (OPIE 1986), but no information on its success there as a biocontrol agent has been found. It seems to have been introduced into Greece at least twice: in 1976 into Corfu (Pappas and Viggiani 1979) and the second time on an unspecified date, with release areas not further specified (Cavalloro and di Martino 1986). Substantial control has been reported from the second introduction (Cavalloro and di Martino 1986).

1.17.3 *Serangium parcesetosum* Sicard (Col., Coccinellidae)

Syn.: *Serangium montazerii* Fursch
Serangium parcesetosum is a predatory beetle native to India (Malausa *et al.* 1988). Known hosts also include four other Aleyrodidae, i.e. *Aleurocanthus woglumi* Ashby, *Aleurolobus barodensis* Maskell, *A. floccosus* (Maskell) and *B. tabaci* (Gennadius) (CABI 2007). Besides Europe, the species was also released against *D. citri* in Turkey, the USA and the former USSR (BIOCAT 2005).

In 1985, *S. parcesetosum* was shipped to France to establish a rearing (Malausa *et al.* 1988). In 1986, 64 adults were released in Corsica. In 1987, 70 adults, and again in 1988, 42 adults were released on the mainland (Malausa *et al.* 1988). The species established successfully, and reductions of pest numbers were reported in the autumn following releases (Malausa *et al.* 1988). However, no recent information on the impact of *S. parcesetosum* on pest numbers has been found.

1.18 *Diaspidiotus perniciosus* (Comstock), San José Scale (Hem., Diaspididae)

Syn.: *Comstockaspis perniciosa* (Comstock); *Quadraspidiotus perniciosus* (Comstock)
Diaspidiotus perniciosus is native to eastern Asia. Today, it is recorded from several countries in Africa, Central, North and South America, Europe and Oceania. In Europe, *D. perniciosus* seems to have appeared first in Hungary in 1928, from where it subsequently spread to other countries. Today, it is recorded from

21 countries, including Switzerland. The most important means of spread is through the transport of infested nursery material.

Diaspidiotus perniciosus is a highly polyphagous species, feeding on fruits, leaves and stems. It is mainly a pest on deciduous fruit trees, including apple, pear, peach, plum and currants, and on woody ornamental plants, and local outbreaks have been observed in different parts of the world. Attacked trees lose their vigour and have a reduced lifespan. *Diaspidiotus perniciosus* has been removed from the quarantine list of the European Union because of its presence in most of the member countries.

Fumigation of seedlings to prevent early infestation and dormant oil spraying for the treatment of plantations are effective chemical control approaches. Crawlers of *D. perniciosus* are affected by most insecticides used against other pests (organophosphates, carbamates, pyrethroids).

There have been many successful special biological control projects against *D. perniciosus* in different parts of the world, especially in the USA, Europe and Russia. Natural enemies recorded on *D. perniciosus* include 38 parasitoid species from 15 genera (particularly from *Aphytis* and *Encarsia*; Hymenoptera) and 24 predatory species from 13 genera (many *Coccinellidae*).

1.18.1 *Chilocorus similis* (Rossi), Asiatic ladybird (Col., Coccinellidae)

Chilocorus similis is a predatory beetle native to Asia (Clausen 1978). Known hosts include four scale species (CABI 2007). Besides Europe, *C. similis* has been introduced into the USA against *D. perniciosus* (BIOCAT 2005). It was also released against other scale insects in Peru and the USA (BIOCAT 2005).

Specimens originating from China were introduced into former Czechoslovakia in 1956. While Clausen (1978) reported that the species did not establish because it was unable to overwinter, other sources report it as present in both the Czech and Slovak Republics (CABI 2007). No information on its success as a biocontrol agent was found.

1.18.2 *Encarsia perniciosi* (Tower) (Hym., Aphelinidae)

Syn.: *Prospaltella perniciosi* Tower

Encarsia perniciosi is a solitary endoparasitoid native to North America (Polaszek *et al.* 1999). Known hosts include seven genera in the family Diaspididae (Heraty *et al.* 2007). Besides Europe, the species was introduced into Australia, Chile, India, New Zealand, Pakistan and South Africa to control *D. perniciosus*. *Encarsia perniciosi* was also used as a biological control agent against two other Diaspididae pests, *A. aurantii* Maskell and *P. pentagona* in Bermuda, New Zealand, Pakistan, South Africa and the USA (BIOCAT 2005). In Europe, the species was introduced to control *C. dictyospermi* in France and Greece (Greathead 1976).

It is recorded as present in Albania, Austria, Bulgaria, former Czechoslovakia, Denmark, France, Germany, Greece, Hungary, Italy, Portugal, Romania, Spain, Switzerland, the former USSR and former Yugoslavia (Serbia and Montenegro) (Noyes 2002; Rasplus *et al.* 2010).

Encarsia perniciosi originating from the USA was probably first introduced into Italy in 1932 (Russo 1951). Additional introductions are thought to have happened in 1946 and/or 1947 (Greathead 1976), but both attempts were apparently unsuccessful (Clausen 1978). In 1958, and again in 1967, a Russian strain obtained via France was released in the Po Valley, Italy, where it established and achieved 60% parasitism within a year after release (Gambaro 1965). Additional releases followed, and overall 'some millions' of *E. perniciosi* were released in a single region (i.e. Valtellina). Releases resulted in a progressive reduction in the level of *D. perniciosus* populations (CABI 2007).

In 1950, *E. perniciosi* originating from the USA were first introduced into Germany to establish a rearing (Klett 1959). A first release of 10,000 individuals followed in 1954 in the Heidelberg area. In 1955, a further 500,000 individuals were released in the same area, and by 1957 parasitism in the release area had reached 40–70%. In 1958/59, some 1,328,000 of a Chinese strain and 10,450 of a Russian strain were again released near Heidelberg (Greathead 1976). In 1960, another 1.5 million *E. perniciosi* were released around Heidelberg and Heilbronn, reaching parasitism rates of up to 85%. Additional collections of scales parasitized by a bisexual race were made in the USA in 1961 and 1962. Overall, from the beginning of 1954 up to 1973, about 27 million individuals were released in Baden Württemberg, Rhine Palatinate and the Plain of the Rhine (Greathead 1976).

In 1951, 40 apple twigs bearing parasitized *D. perniciosus* from the former USSR and Bulgaria were planted under tent-covered, infested trees in former Czechoslovakia (Clausen 1978). In 1953, additional material was received from the former USSR, and in 1956 from China on infested apples (Clausen 1978). In 1958, *E. perniciosi* originating from the USA was received via Germany (Clausen 1978). *Encarsia perniciosi* established permanently in former Czechoslovakia and spread (Clausen 1978), but no information on its success as a biocontrol agent was found.

In 1951, *E. perniciosi*, together with *E. fasciata* (Section 2.2.1), was first introduced for mass rearing in France (Bénassy and Burgerjon 1955). In 1958, field tests were set up in apple and pear orchards near Lyon. By 1959, *E. perniciosi* had spread throughout the orchards, reaching a parasitism rate of 15% (Biliotti *et al.* 1960). Releases in other regions of France began in the 1960s. At some release sites, parasitism rates reached 50% or more; at a site near Lyon, releases in 1964 resulted in the control of *D. perniciosus* by December 1965 (Bénassy *et al.* 1968).

Encarsia perniciosi was introduced into Switzerland in 1956, mass-reared on infested watermelons and *Prunus padus* L. and subsequently released in 1957 at multiple sites. The pest was eradicated in the Geneva, Neuchâtel and Vaud cantons, and greatly reduced in Vallais and Tessin (Mathys and Guignard 1965). In 1958, additional strains of *E. perniciosi* were introduced and reared, and 100,000 individuals were released in five sites. Parasitism of 40% was achieved by the most successful strains from China and Russia. In 1962, a year with particularly high pest numbers, additional releases in Valais and Tessin resulted in 90% parasitism in the release areas (Mathys and Guignard 1965). Interestingly, *E. perniciosi* was able to adapt to new climatic conditions. In eastern Russia, the species has two bisexual generations per year, whereas the pest *D. perniciosus* has one generation per year. In Switzerland, *D. perniciosus* has two and a partial third generation and the Russian strain of *E. perniciosi* has four parthenogenetic generations per year (Greathead 1976).

In 1958, *E. perniciosi* originating from France was imported into the former Yugoslavia for laboratory rearing. It was released in 1960 in 11 different locations in Serbia, Croatia, Slovenia and Bosnia and Herzegovina (Tadić 1967), but no information on its success as a biocontrol agent was found. In 1964, *E. perniciosi* originating from Germany and former Czechoslovakia was imported into Austria and subsequently reared for releases against *D. perniciosus* and *Quadraspidiotus ostraeformis* (CURTIS) in the same year (Argyriou 1981). *Diaspidiotus perniciosus* was reported as having some impact on pest numbers; in 1973, the highest rates of parasitism were 50–60% (Clausen 1978). In 1968, 4000 *E. perniciosi* specimens originating from France were introduced into Greece (Argyriou 1981), and additional imports and releases of 2000 individuals from the USA followed in 1969 and 1977 (Argyriou 1981). The species established permanently but reached only low levels of parasitism, probably due to extensive spraying carried out against other pests in fruit orchards (Argyriou 1981).

In 1969, *E. perniciosi* originating from France was introduced in Spain, mass reared and released with 'good results' (Greathead 1976).

1.18.3 *Rhyzobius lophanthae* (BLAISDELL), scale-eating ladybird (Col., Coccinellidae)

Syn.: *Lindorus lophanthae* (BLAISDELL)

Rhyzobius lophanthae is a predatory beetle native to Australia (Clausen 1978). Its host range includes 23 scale insect species, mainly Diaspididae but also Coccidae and Pseudococcidae (CABI 2007). Besides Europe, *R. lophanthae* was introduced into the former USSR against *D. perniciosus* (BIOCAT 2005). It was also released against other scale insects in Agelega Island, Bermuda, Chile, Cook Islands, India, Kenya, Marianas (Guam), Peru, Philippines, St Vincent, South Africa, Tanzania, Trinidad, Uganda and the USA (BIOCAT 2005). In Europe, the species was also introduced against *P. pentagona* in Italy (Section 1.47.4). It is reported today from Albania, Cyprus, France, Germany, Greece, Italy, Malta, Portugal, Spain and the UK (CABI 2007; Roy and Migeon 2010). Its presence in France is considered a result of natural dispersion (Malausa *et al.* 2008).

In 1956/57, *R. lophanthae* originating from the former USSR was introduced into former Czechoslovakia; the species was unable to overwinter and failed to establish in this country (Clausen 1978).

1.19 *Diuraphis noxia* (Kurdjumov), Russian Wheat Aphid (Hem., Aphididae)

Diuraphis noxia is indigenous to southern Russia, Iran, Afghanistan and countries bordering the Mediterranean Sea (El Bouhssini *et al.* 2011). It was introduced to many other countries worldwide in Africa, Asia, Europe, North and South America. *Diuraphis noxia* is a pest on wheat, and although adults are very unlikely to be transported alive on harvested grains or fodder, overwintering eggs can be transported in this way. *Diuraphis noxia* has spread from eastern or south-eastern to central Europe since 1989, and today it is recorded from Bulgaria, Czech Republic, France, Greece, Hungary Italy, Former Yugoslav Republic of Macedonia, Moldova, Montenegro, Portugal, Serbia, Spain, Ukraine and the UK.

The host range of *D. noxia* appears to be restricted to grasses; it reproduces best and does most damage on cool-season grasses. Major hosts include *Hordeum vulgare* L. (barley), *Triticum aestivum* L. (wheat) and *Triticum turgidum* Desf. (durum wheat). It is also recorded from uncultivated, wild growing Poaceae. *Diuraphis noxia* colonies are found most frequently on the youngest leaves or on freshly emerged flowers and on seedheads. Besides direct feeding damage, the species appears to inject a polypeptide toxin that affects the entire plant. Also, *D. noxia* can cause major indirect losses as a result of transmission of barley yellow dwarf luteovirus. However, in Eurasia, it is only occasionally a serious pest, with short outbreaks being reported. Nevertheless, yield losses of up to 75% are reported in some years from the Crimea.

Because *D. noxia* populations must survive the period between the harvest of one year's crop and the planting of the next, management of crop residues and alternative host plants appear promising.

Chemical control has been and is still used against *D. noxia*, but variation in susceptibility to insecticides suggests that the species may have developed resistance to some. Studies in 1995 revealed, however, that several insecticides currently not in use against this species were effective for its control and only moderately toxic to natural enemies.

Natural enemies recorded on *D. noxia* include 24 parasitoid species from eight genera (many *Aphidius* and *Aphelinus* species), ten pathogen species from seven genera and 65 predatory species from 33 genera. Biological control has been used against this pest in the USA, and four parasitoid species are known to have established.

1.19.1 *Aphidius colemani* Viereck (Hym., Braconidae)

Aphidius colemani is a solitary endoparasitoid of aphids that was thought initially to originate from India, but reclassification of two other species as junior synonyms of *A. colemani* extended its native range to Central Asia and to the Mediterranean (CABI 2007). This endoparasitoid is recorded from 76 Aphididae species from 32 genera (Yu *et al.* 2005). *Aphidius colemani* adapted to *D. noxia* in the course of its spread over North and South America (Starý 1999). Primary hosts are *M. persicae*, *Myzus nicotianae* Blackman and *A. gossypii*, against which it has been particularly effective as a biocontrol agent (CABI 2007). The species is otherwise used as a commercial biocontrol agent in glasshouses, targeting primarily *M. persicae*. *Aphidius colemani* is likely to have escaped from greenhouses; for instance, it escaped from greenhouses in Germany accidentally and subsequently established in cereal fields on *R. padi* L. (Starý 2002). Besides Europe, the species was introduced into Australia, Africa, Brazil, Central America, Tonga and the USA against other aphid species (BIOCAT 2005; CABI 2007). In Europe, *A. colemani* was also introduced against *A. spiraephaga* and *M. donacis* (Sections 1.6.1, 1.32.1). To date, it is present in Europe in Albania, Austria, Belgium, the Czech Republic, Denmark, Finland, France, Germany, Greece, Hungary, Ireland, Italy, Lithuania, Malta, the Netherlands, Norway, Poland, Portugal, Slovak Republic, Spain, Sweden, Switzerland and the UK (Yu *et al.* 2005; CABI 2007; Rasplus *et al.* 2010).

Aphidius colemani originating from Chile was introduced into the Czech Republic in the 1990s and reared on *A. fabae* Scopoli and *R. padi* (Starý 1999). In 1996 and 1997, 10,000 and 50,000 adults were released, respectively, with release sizes of between 200 and 2000 adults per site (Starý 1999). Adults made up only a small number of the released agents, as most of the releases were done with banker units (i.e. beans infested by parasitized *M. persicae* and wheat infested by parasitized *R. padi*), which yielded some hundreds to thousands of adult parasitoids in the field (Starý 1999). Further releases in the Czech Republic were also made against other aphid species, i.e. *A. spiraephaga* Müller, *M. donacis* (Passerini) and *Aphis viburni* Scopoli (Starý and Havelka 2008). Permanent establishment of *A. colemani* in the Czech Republichas been reported (Starý 2002), but no information on its success as biocontrol agent has been found.

Rearings from non-target aphid species in the area in the Czech Republic where *A. colemani* was released indicate that it also attacks at least six additional aphid species, including five 'non-target hosts', i.e. aphid species on non-cultivated plant species (Starý 1999).

1.20 *Dryocosmus kuriphilus* Yasumatsy, Chestnut Gall Wasp (Hym., Cynipidae)

Dryocosmus kuriphilus was first recorded in Europe in 2002 in northern Italy (Brussino *et al.* 2002). Most probably, the pest was introduced as a contaminant of nursery material from China. The chestnut gall wasp has since spread to other parts of Europe, including Croatia, France, Hungary, Slovenia and Switzerland (Gibbs *et al.* 2011).

Dryocosmus kuriphilus causes the formation of galls on buds of chestnut trees, *Castanea* spp., in early summer. These galls reduce the total photosynthetic area, suppress nut production and cause a gradual decline in the vigour of these long-lived and slow-growing trees. Because of the danger it poses, it was added to the European Plant Protection Organization (EPPO) A2 Action list (EPPO 2005).

Since the larval and pupal stages of *D. kuriphilus* are protected inside their galls, conventional chemical control is regarded as largely ineffective. Developing resistant varieties of *Castanea* spp. could potentially be a viable management option, but this will only be beneficial for new plantings and will not help existing chestnut populations (Gibbs *et al.* 2011).

In Europe, numerous gall wasp parasitoids from four families have been reared consistently from *D. kuriphilus* (Gibbs *et al.* 2011).

1.20.1 *Torymus sinensis* Kamijo (Hym., Torymidae)

Torymus sinensis is an ectoparasitoid native to China (Quacchia *et al.* 2008) and is the only parasitoid of *D. kuriphilus* that is considered to be host specific (Gibbs *et al.* 2011). The taxonomic status of this species is uncertain, since it is reported to hybridize with closely related species in its native range (Quacchia *et al.* 2008). This is noteworthy, since three *Torymus* species reared from *D. kuriphilus* galls from Europe (*T. auratus* (Muller), *T. flavipes* Walker and *T. scutellaris* Walker) appear to have the potential to hybridize with *T. sinensis* (Gibbs *et al.* 2011).

Torymus sinensis is univoltine; adults emerge from withered galls in early spring, and females lay eggs into newly formed galls, either on to the body surface of the host larva or on the wall of the larval chamber. *Torymus sinensis* was introduced into Japan as a biological control agent against *D. kuriphilus*, where it decreased *D. kuriphilus* densities below the economic threshold within a few years after its first release (Quacchia *et al.* 2008).

In 2005, adults emerging from field collections made in Japan were released in the Piedmont area, northern Italy (Quacchia *et al.* 2008). Releases continued for several years and were extended to other regions of northern Italy, i.e. Lombardy, Liguria and Latium (Aebi *et al.* 2011). To date, *T. sinensis* has established well at several of the release sites, but according to Gibbs *et al.* (2011) it is too early to assess its effectiveness for the long-term management of the Italian *D. kuriphilus* populations.

1.21 *Eriosoma lanigerum* (Hausmann), Woolly Aphid (Hem., Aphididae)

Eriosoma lanigerum probably originates from eastern North America. The species has spread to many other countries, mainly on apple rootstocks, and now has a worldwide distribution.

Eriosoma lanigerum was first detected in Europe in London in 1787 and spread rapidly to wherever apples were grown. Today, it is recorded from all European countries but Estonia, Iceland, Liechtenstein, Lithuania, Luxembourg, Norway and Slovenia.

Besides apple (*Malus* spp.), *E. lanigerum* is occasionally found on certain other woody host plants in the family Rosaceae, including *Crataegus*, *Sorbus* and

Cotoneaster. In its native range in North America, *Ulmus americana* L. (Ulmaceae) serves as a winter host to *E. lanigerum*. In Europe, however, it remains on apple throughout the year.

Eriosoma lanigerum is an important economic pest of apple. In West Virginia, USA, average yield losses were estimated at 2.4 kg/tree (13 apples), representing a gross loss of US$465.18/ha. *Eriosoma lanigerum* feeds on a variety of plant parts, including stems and roots. Damage is caused mainly by direct feeding and is particularly severe in young trees. Heavily infested trees lack vigour, due to disturbances in nutrient balance. Heavy infestations may also result in the formation of swellings and wounds that permit the entry of cancer-producing fungi. In addition, fruit can be directly affected through the deposition of honeydew from colonies feeding on adjacent branches, which causes cosmetic damage and leads to the growth of sooty moulds on the apples.

Chemical treatment (yellow mineral oil) of the aerial parts of trees has been used in the winter against hibernating immature forms of the aphid. During the growing period, systemic products (in particular vamidothion) can be applied. However, residues may persist on fruit after harvest if sprayed late in the season. Some resistance of *E. lanigerum* to vamidothion has been reported. In nurseries, root dips of fenitrothion or dichlorvos, or soil applications of phorate granules during the spring, can provide effective control. Subterranean populations in orchards can be treated with granules of dimethoate, aldicarb or carbofuran. Spraying with tar distillate increased aphid numbers, probably by destroying the parasitoid *Aphelinus mali* (Haldeman) (Bodo 1934).

Integrated pest management (IPM) has been developed and implemented successfully against *E. lanigerum*, including enhancement of the diversity of arboreal predators and parasitoids, and high populations of key natural enemies, particularly the parasitoid *A. mali*. In addition, four Coccinellidae (Coleoptera) species have been introduced as biological control agents in countries outside Europe (BIOCAT 2005).

Six parasitoid, one pathogen and 73 predatory species belonging to five orders and seven families (Coccinellidae, Chrysopidae, Hemerobiidae, Forficulidae, Lygaeidae, Syrphidae and Cecidomyiidae) have been reported to feed on *E. lanigerum*.

1.21.1 *Aphelinus mali* (Haldeman), woolly aphid parasite (Hym., Aphelinidae)

Aphelinus mali is an endoparasitoid, native to North America (CABI 2007). It has been recorded from 20 aphid host species, but also from two Diaspididae (Hemiptera) and two Lepidoptera species (Noyes 2002). However, *E. lanigerum* is the main host of *A. mali*, and while all stages can be attacked, it prefers third-instar nymphs (CABI 2007). Besides Europe, the species was introduced into several countries in Africa, Central and South America, Australia and New Zealand as a biological control agent against *E. lanigerum* (BIOCAT 2005). *Aphelinus mali* was further introduced into the Cook Islands against *Pseudococcus* spp. (Homoptera). Besides the countries mentioned below, it is also considered as present in Europe in Albania, Bulgaria, the Czech Republic, Hungary, Moldovia, Romania, the Slovak Republic, Slovenia and the former USSR (CABI 2007; Rasplus *et al.* 2010).

In 1920, a first successful shipment of *A. mali* was made from the USA to France, and the 80 parasitoids obtained were used to establish a rearing (Marchal 1929). First releases into the field followed presumably in the same year and were continued in following years. Until 1950, a rearing was kept to supply requests (Pussard 1950). Best results were obtained in the south, but establishling *A. mali* in the northern, wetter parts of France proved to be difficult. Bonnemaison (1965) concluded that premature emergence and a lower fecundity and growth rate compared to its host did not allow for a sufficiently rapid build-up of the parasitoid population density.

Shipments of *A. mali* to Italy were made from France in 1921 and from Uruguay in 1922 (Howard 1929). Adults from both shipments were released in Florence, where they multiplied rapidly and were used as a source of material for further distribution throughout northern and central Italy, and later on also to southern Italy (Howard 1929). Satisfactory control of *E. lanigerum* was achieved at various release sites, but after the Second World War, *A. mali* populations temporarily decreased and *E. lanigerum* populations experienced a resurgence due to the spraying of organophosphates and DDT against *C. pomonella* in apple orchards (Principi *et al.* 1974).

In 1922, *A. mali* was imported into Switzerland from France for propagation (Faes 1928). From 1922 onwards, the parasitoid was released in different localities of the Cantons of Vaud, Valais and Geneva. Subsequentlly, the parasitoid was also released into the German-speaking part of Switzerland (Menzel 1931). Greathead (1976) reported that in Switzerland the woolly aphid and *A. mali* tended to occur at low densities, but that in the 1970s heavier attack rates by the woolly aphid were observed in the French-speaking part of Switzerland, with simultaneously increasing parasitism rates of *A. mali*.

In 1922, *A. mali* was also sent to Belgium, from where excellent results were reported (Marchal 1929). In Portugal, *A. mali* from France was released between 1922 and 1927 (Marchal 1929), and *A. mali* from Spain was released between 1929 and 1933 (Nonell Comas 1938). No information on the impact on the pest species is reported.

In Germany, the first imports of *A. mali* were made from Uruguay in 1923, and the first releases were carried out in 1924. Additional releases of specimens from Italy followed between 1926 and 1933. *Aphelinus mali* contributed substantially to the successful control of *E. lanigerum*; however, complete control was not obtained due to the late appearance of the parasitoid in spring and insufficiently rapid population build-up. Moreover, winter sprays interfered with the parasitoid and required local reintroduction of this species (Sachtleben 1941).

In 1924, first releases of *A. mali* obtained from France were made in the UK (Greathead 1976). Additional releases were made in the 1920s and 1930s, but the species established only temporarily. It was concluded that the use of *A. mali* in the

UK was limited by the climate and poor synchronization with its host (Thompson 1930). However, in 1942, the establishment of *A. mail* was confirmed in Essex and Kent, and complete control was reported at some of the release sites (Greathead 1976). Cold storage to make releases during spring to boost the populations of *A. mali* resulted in successful control of *E. lanigerum* under commercial conditions. Nevertheless, control of *E. lanigerum* by *A. mali* has not been as complete in the UK as in many other European countries (Greathead 1976).

In 1924, *A. mali* was imported from France into the Netherlands, where it was first reared and subsequently released in 80 localities in 1926. Parasite densities were first reported to be low, but by the mid-1930s, *A. mali* was found to be much more abundant and widespread (Schoevers 1934). Detailed investigations by Evenhuis (1962) revealed that in spring *A. mali* was unable to find sufficient hosts, leading to decreased parasitization rates in early summer. In late summer, parasitization increased again, because the rate of reproduction of *E. lanigerum* slowed down.

In 1926, *A. mali* from Italy and Uruguay was imported into Spain for rearing and for releases from 1927 onwards. Breeding and distribution continued until the early 1940s, and where *A. mali* became established, it permanently reduced wooly aphid infestations (Spain 1941). Between 1926 and 1940, releases of *A. mali* were also made in Austria, Cyprus, Denmark, Italy, Malta, Poland and Sweden; siginficant impacts on pest numbers were reported from most of these countries. In Cyprus and Malta, *A. mali* apparently found optimal conditions and completely controlled *E. lanigerum*. In Denmark and Sweden, however, *A. mali* did not always reduce pest populations markedly, because *E. lanigerum* populations were able to increase in spite of parasitism under the favourable weather conditions of late summer and autumn (Greathead 1976).

In general, *A. mali* only attacks aphids above ground. Consequently, where above-ground infestations dominate, control of *E. lanigerum* has been very successful, but where root-feeding populations are important, the results have been less impressive (CABI 2007).

1.22 *Ectomyelois ceratoniae* (Zeller), Locust Bean Moth (Lep., Pyralidae)

Syn.: *Apomyelois ceratoniae* (Zeller)
Ectomyelois ceratoniae is considered to be of Mediterranean origin (Heinrich 1956), but now has a nearly cosmopolitan distribution since it has been introduced into many regions of the world. In Europe, it is currently recorded from Cyprus, France and the former USSR.

Ectomyelois ceratoniae is a highly polyphagous species that attacks fruits of several host plants both during storage and in the field. It is considered a major pest on *Citrus*, date, almond and pomegranate (Mozaffarian *et al.* 2007). The host range includes other crop plants, incuding pigeon pea (*Cajanus cajan*) and Indian tamarind (*Tamarindus indica*).

Fumigation of harvested and stored fruits has been recommended. Chemical treatments used to be applied, but some treatments have since been banned (e.g. the use of organochlorine insecticides in date production; Vreysen *et al.* 2006). Spraying with *B. thuringiensis* is currently used to control *E. ceratoniae*, and research has also begun on the use of the sterile insect technique (Vreysen *et al.* 2006).

Mechanical control methods include physically protecting fruits with nets and plastic bags, cleaning plantations of wind-fallen fruits (reducing overwintering populations) and removing fruit parts used as oviposition sites (Mozaffarian *et al.* 2007). Pheromone traps are also available for this species.

Natural enemies recorded on *E. ceratoniae* include 11 parasitoid species of ten genera and two predator species (CABI 2007). Biological control using *Trichogramma* wasps is applied in Iran (references in Mozaffarian *et al.* 2007).

1.22.1 *Phanerotoma flavitestacea* Fischer (Hym., Braconidae)

Syn.: *Leucobasis flavitestacea* Fischer
Phanerotoma flavitestacea is an endoparasitoid native to the Middle East (Katsoyannos *et al.* 1997). Known hosts include six other Lepidoptera species, *Amyelois transitella* (Walker), *Apomyelois decolor* (Zeller), *Cadra calidella* (Guenée), *Ephestia kuehniella* Keller, *Maruca vitrata* (Fabricius) (Pyralidae/Crambidae) and *Galleria mellonella* (L.) (Galleriidae) (Yu *et al.* 2005; CABI 2007). Besides Europe, the species has also been introduced into California, USA, against *A. transitella*, a pest of *Citrus* (BIOCAT 2005). In Europe, *P. flavitestacea* is reported from Croatia and France (Yu *et al.* 2005).

In 1967, *P. flavitestacea* from Algeria was introduced into France. The parasitoid established there and some impact on pest numbers was recorded (Malausa *et al.* 2008).

1.23 *Gonipterus scutellatus* Gyllenhal, Eucalyptus Weevil (Col., Curculionidae)

Gonipterus scutellatus is native to Australia and has been introduced into Africa, Europe and New Zealand, as well as North and South America. In Europe, it was first recorded in Italy in 1976, and then in France (1978) and Spain (1991). Today, it is also recorded from Croatia, Greece and Portugal. Adults, larvae and eggs may be carried on plant parts used for vegetative propagation, and on seedlings. The larvae and pupae could also be present in the potting soil.

Gonipterus scutellatus is specific to *Eucalyptus* species. *Pinus patula* Schltdl. and Cham is an alternative host for some adults during the winter, but they usually hibernate under loose bark on *Eucalyptus* trunks. In the invaded range, the species is an important defoliator of *Eucalyptus* trees. Feeding damage reduces growth rate and tree vigour and causes the loss of apical dominance. Trees also become more susceptible to attack by other organisms.

Chemical treatment can be applied, but is not recommended because of the disruption it causes to the environment, and also because it could affect honeybees, which are attracted by the long flowering period of *Eucalyptus* trees.

1.23.1 *Anaphes nitens* (Girault) (Hym., Mymaridae)

Anaphes nitens is an egg parasitoid native to Australia (BIOCAT 2005). Known hosts include two Curculionidae, *G. scutellatus* and *Gonipterus gibberus* Boisduval (Noyes 2016). The species was also introduced into Argentina, Brazil, Chile, Kenya, Madagascar, Malawi, Mauritius, New Zealand, St Helena, South Africa, Tanzania, Uganda, Uruguay, the USA and Zimbabwe against *G. scutellatus* and/or *G. gibberus* (BIOCAT 2005). In Europe, it is recorded from France, Italy, Portugal and Spain (Rasplus *et al.* 2010).

In 1978, *A. nitens* was introduced into Italy, from where substantial control has been reported (CABI 2007). In 1979 and 1981, *A. nitens* from Australia was released in France, which resulted in some impact on pest numbers (OPIE 1986; Malausa *et al.* 2008). Releases in Spain started in 1993 and quickly led to excellent control of *G. scutellatus* (Mansilla and Pérez Otero 1996).

1.24 *Heliothrips haemorrhoidalis* (Bouché) (Thys., Thripidae)

The thrips *Heliothrips haemorrhoidalis* is a polyphagous cosmopolitan pest of cultivated ornamental plants and of fruit crops. Its native range is considered to be South America (CSIRO 2009). High infestations cause severe defoliation and, eventually, the death of the plants. This species is uniparental and reproduces by thelytokous parthenogenesis.

Chemical control of this pest can be achieved by using systemic and translaminar insecticides, which are highly toxic and persistent, and thus unsuitable for public parks and gardens (Bernardo *et al.* 2005).

In Europe, two egg parasitoids of *H. haemorrhoidalis* have recently been recorded from southern Italy, namely *Megaphragma mymaripenne* Timberlake and *Megaphragma amalphitanum* Viggiani (Hym., Trichogrammatidae) (Bernardo *et al.* 2005). However, neither of the two appears to have a significant impact on the ability of the pest to cause damage.

1.24.1 *Thripobius javae* (Girault) (Hym., Eulophidae)

Syn.: *Thripobius semiluteus* (Bouček)

Thripobius javae is a uniparental, solitary endoparasitoid. Parasitized thrips larvae appear swollen and the sides of their body are more parallel than the tapered shape seen in healthy thrips larvae. The parasitoid larva kills the host when the latter reaches the prepupal stage. The parasitoid larva then transforms into a black pupa, which remains cemented to the plant's surface until the emergence of the adult parasitoid (Jamieson *et al.* 2008).

Originating in the Australasian/Oriental region, *T. javae* was introduced into Australia, the USA (in the mid-1980s), Israel and New Zealand (Jamieson *et al.* 2008). Surveys carried out in California indicated that up to 60% of *H. hamorrhoidalis* larvae were parasitized by *T. javae* within 2 years after the release of approximately 11,000 wasps at three sites (McMurtry *et al.* 1992).

In Europe, *T. javae* was introduced into Italy from Israel in 1995. The parasitoid was reared and released at several sites, mainly in southern Italy, where it became established and spread widely (Bernardo *et al.* 2005). Today, the species is also recorded from Belgium, Denmark, France, Germany, Italy and the Netherlands (Rasplus *et al.* 2010).

1.25 *Hyphantria cunea* Drury, Fall Webworm (Lep., Noctuidae)

Hyphantria cunea is of North American origin. The species was introduced into Europe (Hungary, then Austria and former Yugoslavia) during the Second World War and has spread further since then. International trade facilitates movement to new areas since *H. cunea* can be carried on vegetative host-plant material as well as on packing materials and in vehicles. Besides Europe, the species has been introduced into Asia. In Europe, *H. cunea* is recorded from Austria, Bosnia and Herzegovina, Bulgaria, Croatia, the Czech Republic, Denmark, France, Germany, Hungary, Italy, Lithuania, Montenegro, Poland, Romania, Serbia, the Slovak Republic, Slovenia and Switzerland.

Hyphantria cunea attacks a wide range of fruit and forest trees, excluding conifers, and the larvae can defoliate whole trees rapidly. The recorded host range in Europe includes 230 species. Major hosts include *Juglans nigra* L. (black walnut), *Malus domestica* Borkh. (apple), *Prunus avium* L. (sweet cherry), *Prunus cerasus* L. (sour cherry), *Prunus domestica* L. (plum) and *Pyrus communis* L. (European pear). The species is also recorded from wild species, including *Acer* (maples), *Fraxinus excelsior* L. (ash), *Populus* (poplars), *Salix* (willows) and *Tilia cordata* Mill. (small-leaf lime). *Hyphantria cunea* was first recorded in Europe in 1940, and became the object of great concern in the 1950s and 1960s. Today, it is well established in Europe, but is no longer spreading and only locally damaging.

Both mechanical (destroying nests) and chemical means are used for controlling *H. cunea*. Besides parasitoids (see below), biological control includes also the use of *B. thuringiensis* sprays, pathogenic organisms (the fungi *Beauveria bassiana* (Bals.-Criv.) Vuill and *Beauveria globulifera* (Spegazzini)) and entomophagous nematodes (*Steinernema feltiae* (Filipjev)).

Natural enemies recorded on *H. cunea* include 44 parasitoid species from 32 genera, 24 predatory species from 19 genera and 20 pathogen species and subspecies.

While no information was found on the success of the individual agents that have been released in Europe, overall the biological control project against *H. cunea* was rated by Greathead (1976) as a failure. Heavy new pest outbreaks developed in Yugoslavia in 1967 and 1973. The reasons for this might have been the limited number of individuals released for some agents, lack of synchronization with *H. cunea* and the fact that some biological control agents require alternative hosts (Greathead 1976).

1.25.1 *Aleiodes sanctihyacinthi* (Provancher) (Hym., Braconidae)

Syn.: *Aleiodes hyphantriae* (Gahan); *Rogas hyphantriae* Gahan

Aleiodes sanctihyacinthi is native to North America. According to Yu *et al.* (2005), this species lives monophagously on *H. cunea.*

In 1954, and again in 1955, *A. sanctihyacinthi* originating from southern Canada was introduced into former Yugoslavia. A total of 167 individuals were released and, although its establishment was considered to have failed (Greathead 1976), the species has been recorded today from Montenegro and Serbia (CABI 2007). No information on its success as a biocontrol agent was found.

Between 1960 and 1963, *A. sanctihyacinthi* originating from Canada was also introduced into former Czechoslovakia (Clausen 1978). Today, it is recorded from both the Czech and Slovak Republics (CABI 2007), but no information on its success as a biocontrol agent has been found.

1.25.2 *Campoplex validus* (Cresson) (Hym., Ichneumonidae)

Syn.: *Sinophorus validus* (Cresson)

Campoplex validus is native to North America (Greathead 1976). Host records include 33 Lepidoptera species from 11 different families (Yu *et al.* 2005). Besides Europe, *C. validus* was also released against *H. cunea* in the former USSR (BIOCAT 2005).

In the 1950s and 1960s, a total of 731 individuals were released in former Yugoslavia (Tadić 1958). *Campoplex validus* is recorded today from Montenegro and Serbia (CABI 2007), but information on its impact on *H. cunea* has not been found.

Between 1960 and 1963, *C. validus* originating from Canada was also introduced into former Czechoslovakia (Clausen 1978), but the species apparently failed to establish (Yu *et al.* 2005).

1.25.3 *Ceratomegilla maculata* (De Geer), spotted ladybird (Col., Coccinellidae)

Ceratomegilla maculata is a predatory beetle native to North America (Greathead 1976). It is recorded as feeding on aphids, but also on the eggs of *H. cunea* (CABI

2007). *Ceratomegilla maculata* is also recorded as present in Central and South America and in New Zealand, presumably as a result of biological control introductions (CABI 2007).

In 1968, 6720 *C. maculata* originating from southern USA were released into former Yugoslavia (Greathead 1976). Today, the species is recorded from Montenegro and Serbia (CABI 2007). No information on its success as a biocontrol agent was found.

1.25.4 *Cotesia hyphantriae* (Riley) (Hym., Braconidae)

Syn.: *Apanteles hyphantriae* Riley

Cotesia hyphantriae is an egg-larval parasitoid native to North America. Host records for the species include 14 Lepidoptera species from 11 genera, including Galacticidae, Hesperiidae and Noctuidae species (Yu *et al.* 2005). In Europe, *C. hyphantriae* is recorded from Bulgaria, former Czechoslovakia, Germany, Hungary, the Netherlands, Poland, Switzerland and former Yugoslavia (Yu *et al.* 2005).

In 1954/55, a total of 1692 individuals from south-eastern Canada and north-eastern USA were released in former Yugoslavia (Tadić 1958), and additional introductions were made in 1958 and 1959 (Clausen 1978). According to Greathead (1976), permanent establishment of this species most probably did not occur, despite the fact that a small number of *C. hyphantriae* were recovered at one release site.

Between 1960 and 1963, *C. hyphantriae* originating from Canada was also introduced into former Czechoslovakia (Clausen 1978), but no information on its success as a biocontrol agent was found.

A reason for the putative failure of this species as a biological control agent might be that it appears to have two generations per year and an alternative host must therefore be available for the spring generation. Attack on *H. cunea* occurs only in early August (Clausen 1978).

1.25.5 *Hyposoter fugitivus* (Say) (Hym., Ichneumonidae)

Hyposoter fugitivus is an endoparasitoid native to North America (Greathead 1976). The species is recorded from 31 host species from different Lepidoptera families, including Lasiocampidae, Noctuidae and Saturniidae (Yu *et al.* 2005). It has also been released against *H. cunea* in the former USSR (BIOCAT 2005). In Europe, *H. fugitivus* is recorded as present in Bosnia and Herzegovina, Bulgaria, the Czech Republic, Macedonia, Montenegro, Serbia, the Slovak Republic and the former USSR (CABI 2007).

A subspecies of *H. fugitivus*, *Hyposoter fugitivus fugitivus* (Syn.: *Phobocampe fugitivus fugitivus*) was introduced into former Yugoslavia in 1954–1955. A total of 477 individuals were released, but the species failed to establish (Tadić 1958).

Between 1960 and 1963, an unknown subspecies of *H. fugitivus* originating from Canada was introduced into former Czechoslovakia (Clausen 1978), but no information on its success as a biocontrol agent has been found.

1.25.6 *Hyposoter pilosulus* (Provancher) (Hym., Ichneumonidae)

Hyposoter pilosulus is native to North America. Besides *H. cunea*, the species is also recorded from two other Lepidoptera species, *Euchaetes egle* (Drury) and *Spilosoma virginicum* Fabricius (Noctuidae) (Noyes 2016).

Between 1960 and 1963, *H. pilosulus* originating from Canada was introduced into former Czechoslovakia (Clausen 1978). Today, it is recorded from both the Czech and Slovak Republics (CABI 2007), but no information on its success as a biocontrol agent has been found.

1.25.7 *Panzeria ampelus* (Walker) (Dip., Tachinidae)

Syn.: *Mericia ampelus* Walker
Panzeria ampelus is a larval parasitoid native to North America. The species is also recorded from another Noctuidae host, *Mamestra configurata* Walker (Wylie 1977). *Panzeria ampelus* was also released against *H. cunea* in the former USSR (BIOCAT 2005).

In 1954/55, a total of 7298 individuals from south-eastern Canada and north-eastern USA were released into former Yugoslavia, but the species failed to establish (Tadić 1958).

Between 1960 and 1963, *P. ampelus* originating from Canada was also introduced into former Czechoslovakia (Clausen 1978), but no information on its establishment or success as a biocontrol agent was found.

1.25.8 *Meteorus bakeri* Cook and Davis (Hym., Braconidae)

Meteorus bakeri is an endoparasitoid native to North America. Besides *H. cunea*, the species is also recorded from three other Lepidoptera species, i.e. *Dasychira plagiata* Walker, *Lophocampa argentata* (Packard) and *Morrisonia latex* Guenée (Noctuidae) (Yu *et al.* 2005).

In 1954/55, some 602 adults of two different *Meteorus* sp. were released into former Yugoslavia. According to Tadić (1958), *M. bakeri* was one of the two species that were released (see also Section 1.25.9).

Between 1960 and 1963, *M. bakeri* originating from Canada was introduced into former Czechoslovakia (Clausen 1978). Today, it is recorded from both the Czech and Slovak Republics (CABI 2007), but no information on its success as a biocontrol agent has been found.

1.25.9 *Meteorus hyphantriae* Riley (Hym., Braconidae)

Meteorus hyphantriae is an endoparasitoid native to North America; it is recorded from 37 Lepidoptera species (Yu *et al.* 2005).

According to Tadić (1958), *M. hyphantriae* was one of two *Meteorus* species that were released into former Yugoslavia in 1954/55 (see also Section 1.25.8).

Between 1960 and 1963, *M. hyphantriae* originating from Canada was introduced into former Czechoslovakia (Clausen 1978). Today, the species is recorded from both the Czech and Slovak Republics (CABI 2007), but no information on its success as a biocontrol agent was found.

1.25.10 *Podisus maculiventris* (Say), spined soldier bug (Hem., Pentatomidae)

Podisus maculiventris is a generalist predatory bug commonly found in North America. The species feeds mainly on Lepidoptera and Coleoptera, but is also recorded as a predator of aphids (CABI 2007). *Podisus maculiventris* was also introduced into France and Poland against *Leptinotarsa decemlineata* (Section 1.29.7). It is recorded in Europe from the former USSR, Greece, Italy, Montenegro and Serbia (CABI 2007).

In 1968, 205 *P. maculiventris* originating from southern USA were released into former Yugoslavia (Greathead 1976). Today, it is recorded from Montenegro and Serbia; unconfirmed records are also reported from Bosnia and Herzegovina, Croatia and Macedonia (CABI 2007).

1.25.11 *Podisus placidus* Uhler (Hem., Pentatomidae)

Podisus placidus is a generalist predator native to North America. It is also recorded to feed on *G. mellonella* (Pyralidae), *Helicoverpa zea* (Boddie) (Noctuidae) and *Malacosoma americanum* Fabricius (Lasiocampidae) (CABI 2007).

In 1968, 200 *P. placidus* originating from southern USA were released into former Yugoslavia (Greathead 1976). The species is recorded today from Montenegro and Serbia; unconfirmed records are also reported from Bosnia and Herzegovina, Croatia and Macedonia (CABI 2007).

1.26 *Icerya purchasi* Maskell, Cottony Cushion Scale (Hem., Margarodidae)

Icerya purchasi, a native of Australia, was introduced into California accidentally around 1868. At about the same time, the species also reached New Zealand and South Africa, and started to spread widely through most of the tropical and subtropical countries of the world. *Icerya purchasi* is mainly moved around on live plant material, particularly on ornamental plants and fruit trees. In Europe, it is currently recorded from Albania, Cyprus, France, Greece, Italy, Malta, Montenegro, Portugal, Romania, Serbia, Spain and Switzerland. Recently, it has started spreading northwards, being recorded for the first time breeding outdoors in Paris, France, in 2000, and in London, UK, in 2002 (CABI 2007).

Icerya purchasi attacks a wide variety of hosts, especially woody plants. The species is an important pest of *Citrus*, *Acacia* spp., *Casuarina* spp. and *Pittosporum* spp., but it can damage many types of fruit and forest trees as well as ornamental shrubs. Damage to plants is caused mostly by sap depletion, which may lead to wilting, leaf drop, dieback, stunted growth and the shoots drying up and dying. As with most sap-sucking insects, the production of honeydew leads to the growth of sooty mould over the leaf surfaces, which blocks light from the mesophyll, thereby reducing photosynthesis. *Icerya purchasi* is particularly problematic in *Citrus* plantations. For instance, after its introduction into California, USA, in the late 19th century, it was recorded killing even large citrus trees and devastating orchards. The relatively recent establishment of *I. purchasi* in the Galapagos Islands raised environmental concern because the species had also been found to attack some of the endemic flora (CABI 2007).

The insect growth regulator, pyriproxyfen, was found to be as effective in controlling *I. purchasi* as the organophosphorus insecticides, methidathion and omethoate. However, all life stages of *I. purchasi* are covered with wax, which reduces the effectiveness of most chemical insecticides. In addition, the use of insecticides prevents regulation by natural enemies, which has proven to be a highly successful method of control with this species.

Natural enemies recorded on *I. purchasi* include three parasitoid species, i.e. *Cryptochaetum iceryae* (Williston), *Cryptochaetum monophlebi* (Skuse) (Cryptochetidae, Diptera) and *Macrosiagon octomaculatum* (Gerstaecker) (Ripiphoridae, Coleoptera), and 14 predatory species from 11 genera. A predatory beetle from Australia, first introduced into the USA in 1888, was the first ever example of a successful classical biological control project.

1.26.1 *Rodolia cardinalis* (Mulsant), vedelia beetle (Col., Coccinellidae)

Rodolia cardinalis is a highly host-specific predatory beetle native to Australia. Besides *I. purchasi*, it is recorded as attacking only three additional species in the same genus, i.e. *Icerya aegyptiaca* (Douglas), *Icerya montserratensis* Riley and Howard and *Icerya seychellarum* (Westwood) (CABI 2007). In Italy, it has also been reported attacking another Margarodid scale, *Gueriniella serratulae* (F.), but cannot survive on it. In 1888, *R. cardinalis* was imported from Australia into the USA and immediately proved highly effective in controlling its host (CABI 2007). *Rodolia cardinalis* was subsequently released in about 57 countries and has been effective in controlling *I. purchasi* under a wide range of conditions, except in areas where the indiscriminate use of insecticides has killed the biological control agent. In areas with extreme winters, which kill off the populations, periodic reintroduction has been necessary (CABI 2007). Factors contributing to the success of *R. cardinalis* are: (i) six generations in a year compared with three of its host; (ii) a high degree of host specificity; and (iii) the same habitat requirements as its host. Worldwide, the species was also introduced into numerous countries in Africa, Asia, Central, North and South America and New Zealand against *I. aegyptiaca* and *I. seychellarum* (CABI 2007). In Europe, it is currently recorded from Albania, Bosnia and Herzegovina, Croatia, Cyprus, France, Germany, Greece, Italy, Macedonia, Malta, Montenegro, Portugal, Serbia, Spain, Switzerland, the UK and Ukraine (CABI 2007; Roy and Migeon 2010).

In 1897, *R. cardinalis* from the USA was first introduced into Portugal. Within 2 years the heavily infested orchards around Lisbon were virtually cleared from *I. purchasi* (Marchal 1913). At some stage between 1926 and 1950, 200 beetles were again introduced from Spain, presumably as a result of a private request (Gómez Clemente 1954b).

In February 1901, 1 year after *I. purchasi* was detected in Italy, *R. cardinalis* originating from Portugal was sent to Italy, but the beetles did not survive the journey (Greathead 1976). A few months later, a shipment of 36 beetles was obtained from California, and further shipments from Portugal followed. Their numbers were first multiplied and *R. cardinalis* was subsequently released in the field. Further introductions from France in 1921 and 1923 were used to establish rearing cultures and to distribute *R. cardinalis* to various parts of Italy (Priore 1971). Overall, *I. purchasi* was controlled efficiently by *R. cardinalis*, except where low winter temperatures or excessive use of insecticide limited the potential of the biological control agent (Priore 1971).

In 1910/11, *R. cardinalis* originating from Italy was introduced into former Yugoslavia, and gave 'complete control between May and October' (Wahl 1914).

In 1911, *R. cardinalis* originating from Italy was released in Malta (Borg 1932). Until 1921, repeated releases were made from rearing cultures to compensate for heavy winter losses of the beetles in exposed situations. Excellent control was achieved once colonies were established in sheltered sites. Feeding on *I. aegyptiaca*, a species feeding on various Asteraceae species, is thought to have contributed to its successful winter survival (Greathead 1976).

In 1912, a first shipment of eight *R. cardinalis* originating from Italy was made to France, only a few months after *I. purchasi* was first discovered in the Alpes Maritimes (Marchal 1913). A rearing was established from only three surviving beetles. Further supplies from Italy, Portugal and the USA arrived in the same year, and a first release of 100 individuals was made in August 1912. By the end of the year, *I. purchasi* was under control at the first release site (Marchal 1913). By 1930, some 50,000 beetles had been distributed in France and to other countries (see below), always with excellent results (Greathead 1976).

Between 1922 and 1924, various shipments of *R. cardinalis* originating from France, Italy, Portugal and Uruguay were sent to Spain. Between 1926 and 1950, some 134,618 individuals were released in different parts of Spain (Gómez Clemente 1954b). Very effective control was achieved on the Mediterranean coast and in temperate regions inland, but in the north and in harsher inland areas re-colonization by *R. cardinalis* had to be carried out in the spring every year (Gómez Clemente 1954b).

Between 1924 and 1929, *R. cardinalis* originating from France was introduced and released repeatedly in the southern parts of Switzerland. Good control was achieved in the summer months, but apparently the beetle did not establish permanently (Geier and Baggiolini 1950).

In 1927, *R. cardinalis* originating from France and Italy was introduced into Greece, from where good control was reported (Ayouantis 1940).

In 1938, *R. cardinalis* originating from Egypt was released in Cyprus, but failed to establish. It was reintroduced and released in 1939, and this time it established and provided effective control of *I. purchasi* (Wood 1963).

1.27 *Lepidosaphes beckii* (Newman), Purple Scale (Hem., Diaspididae)

Syn.: *Cornuaspis beckii* (Newman)

Lepidosaphes beckii originates from China. Transport of infested plant material greatly favoured the dispersal of this species, and today *L. beckii* is documented as having a worldwide distribution. In Europe, it is listed as being present in Bosnia and Herzegovina, Croatia, Cyprus, France, Greece, Italy, Macedonia, Malta, Montenegro, Portugal, Romania, Serbia and Spain.

Lepidosaphes beckii primarily infests *Citrus* trees. Other known minor host plants are *Agave sisalana* Perrine (sisal hemp), *Elaeagnus* spp., *Mangifera indica* L. (mango) and *Musa* (banana). Heavy infestation by *L. beckii* can also lead to leaf fall, leaf and shoot malformation and, in extreme cases, to the death of the host. Damage to fruit in heavy infestations can result in spotting and deformity of fruits, which affects their market value.

Chemical control can be achieved by spraying infested plants with mineral oils and insecticide (e.g. organophosphates) at critical points during the season. However, *L. beckii* populations can be controlled successfully by natural enemies, and the use of broad-spectrum insecticides can interfere with biological control. Mechanical control can be achieved by scraping and scrubbing to remove scales.

A number of chalcidoid Hymenoptera are important natural enemies. Altogether, known natural enemies include 22 parasitoid species from eight genera (mainly *Aphytis* and *Encarsia* species) and 15 predatory species from 11 genera (several *Chilocorus* species).

1.27.1 *Aphytis coheni* DeBach (Hym., Aphelinidae)

Aphytis coheni is an ectoparasitoid attacking nymphs and adults of *L. beckii*. It was first described from Israel, but it is thought that the species might originally be from south-eastern Asia (Avidov *et al.* 1970). *Aphytis coheni* is also recorded from six other scale insects, i.e. *A. aurantii* Maskell, *A. citrina*, *Aspidiotus hederae* (Vall.), *A. nerii* Bouché, *C. dictyospermi* (Morgan) and *H. lataniae* (Noyes 2002). Besides Europe, this ectoparasitoid was introduced into Australia, South Africa and the USA against another Diaspididae species, *A. aurantii* (BIOCAT 2005). In Europe, *A. coheni* was also introduced into Greece against *C. dictyospermi* (Section 1.11.1).

In 1960, 1000 *A. coheni* were introduced into Cyprus, where the species established (Greathead 1976). No information on its impact on *L. beckii* was found.

1.27.2 *Aphytis lepidosaphes* Compere (Hym., Aphelinidae)

Aphytis lepidosaphes is an ectoparasitoid native to Asia. It was first discovered in China in 1949 (Hill 2008). Besides *L. beckii*, the species is also reported from *Lepidosaphes gloverii* (Packard), *A. aurantii* and *C. aonidum* (Noyes 2002), and has been used as biocontrol agent of *L. beckii* worldwide (CABI 2007). It was first imported into California, USA, from China in 1948 (CABI 2007). Besides Europe, it has since also been introduced into Argentina, Australia, Brazil, Chile, Ecuador, El Salvador, Israel, Mexico, Peru, Puerto Rico, South Africa and the USA, achieving in general excellent control of *L. beckii* (BIOCAT 2005). In Europe, it is reported from Cyprus, France, Greece, Italy (Sicily) and Spain (CABI 2007).

In 1961, *A. lepidosaphes* from California was released in Cyprus, from where partial control of *L. beckii* was reported (CABI 2007). In 1962, 105 *A. lepidosaphes* from the USA were released in Greece. In spite of the small numbers available for release, the species established and provided substantial control (Argyriou 1974). In 1972/73, *A. lepidosaphes* from the USA was released in France, where the species established (Malausa *et al.* 2008); however, no information on the impact of this biocontrol agent was found. In 1976, *A. lepidosaphes* was shipped from France to Spain, from where substantial control of *L. beckii* has been reported (Carrero 1980).

1.27.3 *Chilocorus circumdatus* (Gyllenhal in Schönherr) (Col., Coccinellidae)

Chilocorus circumdatus is a predatory beetle native to China, India, Indonesia and Sri Lanka (Waterhouse and Sands 2001). It is recorded from ten Diaspididae hosts, but the host range is likely to be relatively broad within the family. *Chilocorus circumdatus* was also introduced into South Africa and the USA as a biocontrol agent against *A. aurantii*, into Australia against *Comstockaspis perniciosus* (Comstock), into Bermuda against *P. pentagona* and into Hawaii against unspecified scale insects (BIOCAT 2005). In addition, it was introduced into Australia for the biological control of aphids (BIOCAT 2005).

Between 1968 and 1969, 265 *C. circumdatus* from India were introduced into Cyprus (Greathead 1976). The species established (CABI 2007), but no information on its impact on *L. beckii* was found.

1.27.4 *Chilocorus hauseri* Weise (Col., Coccinellidae)

Chilocorus hauseri is a predatory beetle native to China, India and Myanmar (Poorani 2004). Known hosts include two Diaspididae species, i.e. *L. beckii* and *A. aurantii* (CABI 2007), but its real host range is likely to be wider. The species was also introduced into the USA as a biocontrol agent against another Diaspididae, *A. aurantii* (BIOCAT 2005).

Between 1968 and 1969, 578 *C. hauseri* from India were introduced into Cyprus (Greathead 1976). The species established (CABI 2007), but no information on the impact of this biological control agent on *L. beckii* was found.

1.28 *Lepidosaphes gloverii* (Packard), Glover Scale (Hem., Diaspididae)

Syn.: *Insulaspis gloverii* (Packard)
Lepidosaphes gloverii is presumed to originate from China. Around 1840 it is thought that the glover scale was first imported accidentally into Florida, USA, from where it subsequently spread to many other countries. Today, the species is recorded from countries in Africa, Europe, Central, North and South America, and from Oceania. In Europe, it is listed as being present in France, Greece, Italy and Spain.

The host range of *L. gloverii* includes eight plant families and 19 genera, including *Citrus*, *Codiaeum variegatum* (L.) Rumph. ex A.Juss. (croton) *Fortunella* (kumquats) and *M. indica* (mango). It attacks all *Citrus* cultivars. Heavy infestations of *L. gloverii* can cause a delay in the development of colour in maturing fruit because the area around the scale insect remains green. Another direct result of a heavy infestation is complete yellowing of the leaves, which is followed by leaf drop and twig dieback. Sooty mould can grow on honeydew deposits on leaves.

Chemical control of *L. gloverii* is seldom practised because the species is mostly kept under good biological control, provided that ants are controlled and that chemicals used for the control of other pests do not affect natural enemies.

Natural enemies of *L. gloverii* include two parasitoid species (*Arrhenophagus chionaspidis* Aurivillius and *Encarsia elongata*) and one predator (*Pharoscymnus tetrastictus* Sicard).

1.28.1 *Encarsia brimblecombei* (Girault) (Hym., Aphelinidae)

Syn.: *Encarsia elongata* (Dozier); *Prospaltella elongata* Dozier; *Encarsia herndoni* (Girault)
Encarsia elongata is native to the oriental region (Hayat 1989). Known hosts of *E. elongata* include five Diaspididae host species, i.e. *L. gloverii*, *Lepidosaphes ulmi* (L.), *A. aurantii*, *C. aonidum* and *Fiorinia theae* Green (Noyes 2002). It has also been used

as a biological control agent against *L. gloverii* in the USA (BIOCAT 2005). In Europe, *E. elongata* is recorded from Albania, France, Italy and Spain (Noyes 2002; Rasplus *et al.* 2010).

In 1979, *E. elongata* was introduced into Spain, from where substantial control of *L. gloverii* was reported (Cavalloro and di Martino 1986). According to CABI (2007), its permanent establishment there is uncertain. In 1987, *E. elongata* from Spain was introduced into Corsica (Malausa *et al.* 2008), but no information on its impact on *L. gloverii* was found. According to CABI (2007), the species was also introduced into Italy, but its establishment has not been ascertained.

1.29 *Leptinotarsa decemlineata* Say, Colorado Potato Beetle (Col., Chrysomelidae)

Leptinotarsa decemlineata originates from south-western North America. In 1875, it was first discovered in France, from where it spread eastwards. Today, *L. decemlineata* is distributed over the whole of southern, central and eastern Europe, but it has not yet established in Cyprus, Ireland, Malta, any of the Scandinavian countries or the UK. The species has also spread further eastwards to most potato-growing areas in Turkey and the former USSR. More recently, it has been recorded in western China. In the EPPO region, *L. decemlineata* has not yet reached the extent of its possible geographic distribution. Adults and larvae are transported readily on potato plants and tubers, and it also spreads by its own natural means across land areas. It cannot be stopped by phytosanitary measures; however, its spread has slowed down considerably in recent years due to international collaborative actions.

Leptinotarsa decemlineata attacks Solanaceae, in particular potatoes and various other cultivated crops, including tomatoes and aubergines). It also attacks wild Solanaceae plants, which can act as a reservoir for infestation. The species feeds on leaves, stems and the tubers of its host plants. *Leptinotarsa decemlineata* is one of the most economically damaging insect pests of potato. If plants become entirely defoliated prior to tuber initiation, total crop loss will result. Currently, the cost of controlling *L. decemlineata* infestations in the eastern USA averages between US$138/ha and US$368/ha. The species has developed resistance to many insecticides, and millions of dollars are invested into developing new insecticides and genetically modified crops that produce insecticidal toxins.

Many chemical insecticides have been used to control Colorado potato beetle. Over time, the species has, however, developed resistance to many insecticides. Hence, transgenic potatoes expressing the gene for *B. thuringiensis* subsp. *tenebrionis* Cry3A delta-endotoxin have been approved for commercial use in the USA.

Crop rotation delays the colonization of the crop by overwintering adults and reduces population densities within the crop.

Natural enemies recorded from *L. decemlineata* include 22 parasitoid species from 12 genera (including several *Hexamermis* and *Myiopharus* species), 53 predatory species from 36 genera, 19 pathogen species from ten genera and several *B. thuringiensis* subspecies.

1.29.1 *Horismenus puttleri* Grissell (Hym., Eulophidae)

Syn.: *Edovum puttleri* Grissell
Horismenus puttleri is an egg parasitoid native to South America (Colombia) (Obrycki *et al.* 1985). Known hosts include four other species of the genus *Leptinotarsa* (Noyes 2002). The species was also introduced into the USA and Uzbekistan against *L. decemlineata* (BIOCAT 2005). The ability of augmentative releases of *H. puttleri* to control *L. decemlineata* was assessed in Canada, Italy and the USA. The results indicated that releases of *H. puttleri* were ineffective against first-generation eggs, but had a greater impact on second-generation eggs (CABI 2007); for example, mass releases of *H. puttleri* (47,000 adults) against second-generation eggs of *L. decemlineata* resulted in a total reduction of 50.5% of the pest in the USA (Sears and Boiteau 1989).

Between 1980 and 1988, *H. puttleri* was introduced into Italy, but releases resulted in only a temporary establishment of the species (Maini *et al.* 1990).

1.29.2 *Lebia grandis* Hentz (Col., Carabidae)

Lebia grandis is a predatory beetle native to North America (Weber *et al.* 2006). Adults prey on the eggs and larvae, while the larvae feed on the fully developed larvae and pupae of *L. decemlineata* and the congeneric species *Leptinotarsa juncta* and *Leptinotarsa haldemani* (Weber *et al.* 2006).

In 1934, 3457 *L. grandis* beetles from the USA and Canada were introduced into France, where the species persisted for several years at the single site where it was released (Chaboussou 1938). According to Chaboussou (1939), the French climate over most of the range of the host is presumed to be too dry for the beetle to spread. According to CABI (2007), it is still present in France.

1.29.3 *Myiopharus doryphorae* (Riley) (Dip., Tachinidae)

Syn.: *Doryphorophaga doryphorae* (Riley)
The genus *Myiopharus* (now including *Doryphorophaga*) is a North American genus including 15 species of parasitic flies mainly associated with Chrysomelidae (O'Hara and Mahony 2009b). Some species also develop in other Coleoptera families, i.e. in Coccinellidae and Scarabaeidae, and in some Lepidoptera families (Gelechiidae, Nymphalidae and Tortricidae) (O'Hara and Mahony 2009b). *Myiopharus doryphorae* is one of several species of this genus that are recorded from *L. decemlineata* (O'Hara and Mahony 2009b). Known hosts of *M. doryphorae* include *L. decemlineata* and other Chrysomelidae, but also the Coccinellidae species *Epilachna varivestis* Mulsant and the Lepidoptera species *Nymphalis antiopa* (Linnaeus) (O'Hara and Mahony 2009b). Besides Europe, the species was also introduced into the former USSR against *L. decemlineata* (BIOCAT 2005). In Europe, *M. doryphorae* is recorded from Poland and other East European countries (CABI 2007).

In 1967, *M. doryphorae* was introduced into Poland, where it became established (CABI 2007). No information on its impact as a biocontrol agent was found.

1.29.4 *Myiopharus* sp. (Dip., Tachinidae)

Around 1964, an unspecified *Myiopharus* (formerly *Doryphorophaga*) species from North America was introduced into Italy (Greathead 1976), but it is thought that it did not establish.

1.29.5 *Perillus bioculatus* (Fabricius), two-spotted stink bug (Hem., Pentatomidae)

Syn.: *Perilloides bioculatus* (Fabricius)

Perillus bioculatus is a predatory bug native to North America (Thomas 1994). Known hosts include *L. decemlineata* and the Lepidoptera cabbage pest *Trichoplusia ni* (Noctuidae) (CABI 2007). The species feeds on adults, eggs and larvae of *L. decemlineata* and contributes to its mortality, but natural populations are generally ineffective in suppressing beetle numbers, especially when densities are high (CABI 2007). However, augmentative releases of *P. bioculatus* can control first-generation eggs and larvae of *L. decemlineata* under short-season conditions in Quebec (CABI 2007). According to CABI (2007), *P. bioculatus* has been recorded from Belgium, Bosnia and Herzegovina, Croatia, the Czech Republic, France, Germany, Greece, Italy, the Former Yugoslav Republic of Macedonia, Montenegro, Poland, Romania, Serbia and the Slovak Republic. However, permanent establishment in the wild has only been confirmed from Bulgaria and Greece (Rabitsch 2010; Simov *et al.* 2012).

In 1929, *P. bioculatus* was first introduced into France, but successful breeding was only achieved in 1933 with 248 individuals originating from the USA. An additional shipment was made in 1935 from Canada consisting of 15,000 adults of *P. bioculatus* and *Perillus circumcinctus* Stål (Briand 1936). Releases of these predators were made from breeding cages set up in the field throughout France until 1941, but neither of the two natural enemies became established (Greathead 1976).

From 1956 onwards, breeding stocks of *P. bioculatus* were sent repeatedly from Canada and the USA to the biological control institute in Darmstadt, Germany. In 1957, institutions from 12 European countries participated in a new and well-coordinated programme to establish *P. bioculatus* and *Myiopharus doryphorae* (see Section 1.29.3) in Europe.

Greathead (1976) provides a detailed overview of the various releases made in the different countries. In 1959, *P. bioculatus* was released in Poland, but no information on the impact of the species as a biocontrol agent was found. From 1960 to 1962, *P. bioculatus* was again released in France. The first results of field experiments set up in 1962 were promising, but the species suffered from heavy mortality during the following winter and the viability of eggs from these overwintered adults was poor (Le Berre and Portier 1963). Between 1961 and 1962, *P. bioculatus* was sent from Germany to Belgium. After two unsuccessful attempts, 928 adults were released from a field-cage culture (Moens 1963). No information on its subsequent establishment was found. Between 1964 and 1967,

P. bioculatus from Germany was introduced into Italy. Small numbers were released in 1964 but culture numbers increased over time and, in 1967, 55,000 were released. While the species failed to establish according to Greathead (1976), it is recorded as present according to CABI (2007). Heavy winter mortality is thought to be the main reason for the overall disappointing results of *P. bioculatus* in controlling *L. decemlineata*. In 1963, *P. bioculatus* was released in Germany, but only a few individuals were found the year after release. The failure was attributed to a lack of synchronization between overwintering hosts and the predator. In 1964, 41,800 *P. bioculatus* were introduced into Hungary, but 1 year later no recoveries were made. Again in 1965, 57,600 nymphs were released in Hungary but failed to establish. The failure was attributed to unfavourable weather conditions, i.e. cold winter temperatures (Greathead 1976). From 1965 onwards, releases of *P. bioculatus* were attempted in the Former Yugoslav Republic of Macedonia, but no data on either field releases or success are available. High parasitism rates by a *Trisolcus* (= *Asolcus*) sp. (Hym., Scelionidae) were observed in the breeding facilities and this was considered as a potential serious limiting factor in the establishment of *P. bioculatus* (Greathead 1976). Between 1965 and 1966, *P. bioculatus* from Germany was introduced into former Yugoslavia; no information on the impact of the species as a biocontrol agent was found. In 1966, 43,000 *P. bioculatus* were introduced into former Czechoslovakia. According to Greathead (1976), some individuals were recovered after the first winter, but apparently no permanent establishment was achieved.

In 1966, the *P. bioculatus* programme was reviewed by the study group. It was agreed that every possible effort had been made to establish *P. bioculatus* in Europe and that the activity of the group should be suspended (Franz 1966).

1.29.6 *Perillus circumcinctus* Stål (Hem., Pentatomidae)

Syn.: *Perilloides bioculatus* (Fabricius)
Perillus circumcinctus is a predatory bug native to North America (Thomas 1994). Known hosts include *L. decemlineata*, *Trirhabda* spp. and *Calligrapha californica* (Chrysomelidae) (Evans 1982).

In 1935, a shipment was made from Canada to France containing 15,000 adults of *P. circumcinctus* and *P. bioculatus* (Briand 1936; Section 1.29.5). Releases of these predators were made from breeding cages set up in the field throughout France until 1941, but *P. circumcinctus* did not become established.

1.29.7 *Podisus maculiventris* (Say), spined soldier bug (Hem., Pentatomidae)

Podisus maculiventris is a generalist predatory bug commonly found throughout North America, east of the Rocky Mountains. The species feeds mainly on Lepidoptera and Coleoptera, but is also recorded as a predator of aphids (CABI 2007). *Podisus maculiventris* was also used as a biological control agent against

H. cunea (Lepidoptera, Erebidae) in former Yugoslavia (Section 1.25.10). The species has been recorded from Greece, Italy, Montenegro and Serbia; unconfirmed records are also reported from Bosnia and Herzegovina, Croatia and Former Yugoslav Republic of Macedonia (CABI 2007). However, according to Rabitsch (2010), it has not established permanently in the field in Europe.

Extensive releases of *P. maculiventris* from the USA were made from 1935 onwards in France (Greathead 1976). No information on the impact of the species as a biocontrol agent was found. In the 1980s, seasonal releases of some 40,000–100,000/ha second and third instars of *P. maculiventris* were made along the Black Sea coast in Moldavia, Russia and Ukraine. These resulted in a significant reduction of both egg and larval densities of *L. decemlineata* in potato and aubergine fields, but *P. maculiventris* did not establish permanently (De Clercq *et al.* 1998).

1.30 *Lymantria dispar* (L.), Gypsy Moth (Lep., Erebidae)

Lymantria dispar is native to Europe and Asia. Human activity facilitates the dispersal of the species via *L. dispar* egg masses that have been laid on logs, cars, trucks, trains, boats or containers. The European strain was introduced accidentally in 1869 from France into the USA. The species is also recorded from countries in North Africa (Algeria, Morocco, Tunisia), and the risk for accidental introduction into other countries where it is not yet present is high. In Europe, *L. dispar* occurs in many countries but has not yet been recorded officially from Bosnia and Herzegovina, Estonia, Iceland, Ireland, Latvia, Liechtenstein, Luxembourg, Malta, Norway or Slovenia (CABI 2007).

Lymantria dispar has an exceptionally broad host range; in Eurasia, larvae feed on leaves and infloresences of several hundred hosts. Oak species (*Quercus* spp.) are considered the preferred host tree, but heavy defoliation is also observed on several other tree genera such as *Carpinus*, *Castanea*, *Fagus*, *Populus* and *Salix*. Damage may also be observed on orchard trees such as *Malus* (apple), *Pyrus* (pear) and *Prunus* (stone fruit). Outbreaks usually start on a preferred host, oak for example, but as the population density increases other species are subsequently attacked. Severe defoliation results in reduced growth increment and crown dieback, and occasionally in increased tree mortality. In the native range in Eurasia, outbreaks have been reported but tend to be localized and of short duration. *Lymantria dispar* can be a nuisance in urban environments, where it can defoliate ornamental trees and shrubs. In addition, caterpillar hairs can provoke allergenic reactions.

The control of *L. dispar* outbreaks relies mainly on aerial chemical applications. More recently, biochemical insecticides have replaced broad-spectrum, chemical insecticides. The most favoured bioinsecticide to control *L. dispar* is the bacteria *B. thuringiensis*.

Mechanical egg mass removal may be used in high-value stands such as gardens or recreation areas. Silvicultural manipulations based on thinning strategies to reduce host species numbers have been applied to manage gypsy moth populations in North America, but the most susceptible tree species (i.e. oaks) are also the most valuable timber species.

Overall, over 100 species of parasitoids from 56 genera, 44 predator species from 27 genera, 45 pathogen species from 25 genera and many *B. thuringiensis* subspecies have been reported as natural enemies in Eurasia (CABI 2007). Since its introduction into North America in the 19th century, *L. dispar* has been the target of several extensive biological control programmes, and about 80 species of natural enemies, parasitoids, predators and pathogens have been introduced into North America and elsewhere from 1906 to the present.

1.30.1 *Ooencyrtus kuvanae* (Howard) (Hym., Encyrtidae)

Syn.: *Ooencyrtus kuwanae* (Howard); *Ooencyrtus kuwanai* (Howard)
Ooencyrtus kuvanae is an egg parasitoid that is native to Japan (Brown 1984) and has been reared from various Lepidoptera species (Noyes 2002). In Europe, the species is recorded from Austria, Bosnia and Herzegovina, Bulgaria, Czech Republic, France, Germany, Italy, Moldova, Montenegro, Poland, Portugal, Romania, Russia, Serbia, Slovak Republic, Spain, Switzerland, Ukraine and former Yugoslavia (Noyes 2002; CABI 2007; Rasplus *et al.* 2010). Besides intentional introductions, both active dispersal and passive displacement (e.g. parasitized egg clusters on vehicles) have played a role in the spread of this species across Europe (Greathead 1976).

In 1922, *O. kuvanae* was first introduced into former Czechoslovakia, where the species established (Clausen 1978). No information on its impact as a biocontrol agent was found. *Ooencyrtus kuvanae* was shipped from the USA to Spain in 1923 and cultured for several years. Between 1923 and 1924, and again in 1927 and 1933, *O. kuvanae* was released at various locations in Spain. The species established and had some impact on pest numbers; some 20–25% parasitism was recorded at one of the release sites by 1955/56 (Templado 1957). In 1932, *O. kuvanae* from Morocco was introduced into Portugal, where it established but contributed little to the control of *L. dispar* (Greathead 1976).

In 1958, *O. kuvanae* was found in the Former Yugoslav Republic of Macedonia (Tadić and Bincev 1959). It had probably colonized Macedonia by natural spread from Bulgaria, where the species was known to occur prior to 1957 (Greathead 1976). In 1959, about 10,000 individuals of *O. kuvanae* were released on a row of roadside trees in Yugoslavia; 1 year later, average parasitism of *L. dispar* eggs by *O. kuvanae* had reached about 8% (Tadić 1962). The parasitoid was also released in the Slovak Republic (Capek, personal communication in Greathead 1976), where it established successfully but only reached low densities (Zubrik and Novotny 1997).

1.31 *Megoura viciae* Buckton, Vetch Aphid (Hem., Aphididae)

Megoura viciae is native to northern Europe (Clausen 1978). It is recorded as a relatively unimportant pest of Fabaceae, in particular *Vica* and *Medicago* species (Clausen 1978). The species is also known as a vector for viruses (Saucke *et al.* 2008).

1.31.1 *Aphidius megourae* Starý (Hym., Braconidae)

Aphidius megourae is native to Russia (Yu *et al.* 2005). It is recorded from four Aphididae species, i.e. *A. pisum*, *A. craccivora*, *A. fabae* and *M. viciae* (Yu *et al.* 2005). In Europe, the species is recorded from the Czech Republic, France, Hungary and Poland (Yu *et al.* 2005).

In 1962, *A. megourae* was introduced into former Czechoslovakia (Clausen 1978), where it established (Yu *et al.* 2005). However, no information on its impact as a biocontrol agent was found.

1.32 *Melanaphis donacis* (Passerini) (Hem., Aphididae)

Melanaphis donacis is native to Europe and has been introduced accidentally into the USA (Dudley *et al.* 2008). *Melanaphis donacis* is recorded mainly from *Arundo donax* L., a giant grass native to the Mediterranean region that has long been cultivated throughout the world for use in making mats, roofing material, erosion control and as an ornamental (Starý 1976). This plant species is now considered invasive in some areas such as the southern USA (Lambert *et al.* 2010).

No further information on the problems caused by *M. donacis* on *A. donax* in former Czechoslovakia was found, but we assume that the economic value of *A. donax* motivated attempts to control the aphid there.

1.32.1 *Aphidius colemani* Viereck (Hym., Braconidae)

Aphidius colemani is a solitary endoparasitoid of aphids that was initially thought to originate from India, but reclassification of two other species as junior synonyms of *A. colemani* extended its geographical range from central Asia to the Mediterranean (CABI 2007). It is recorded from 76 host species from 32 genera (Yu *et al.* 2005). Primary hosts are *M. persicae*, *M. nicotianae* Blackman and *A. gossypii* (CABI 2007). The species is used mainly as a commercial biocontrol agent in greenhouses, targeting primarily *M. persicae*. *Aphidius colemani* has escaped from greenhouses, such as in Germany, where it subsequently became established in cereal fields on *R. padi* (Starý 2002). Besides Europe, the species has been introduced into Africa, Australia, Brazil, Central America, Tonga and the USA against other aphid species (BIOCAT 2005; CABI 2007). In Europe, *A. colemani* has also been introduced against *A. spiraephaga* and *D. noxia* (Sections 1.6.1, 1.19.1). It is considered as established outside greenhouses in Albania, Austria, Belgium, the Czech Republic, Denmark, Finland, France, Germany, Greece, Hungary, Ireland, Italy, Lithuania, Malta, the Netherlands, Norway, Poland, Portugal, the Slovak Republic, Spain, Sweden, Switzerland and the UK (Yu *et al.* 2005; CABI 2007; Rasplus *et al.* 2010).

In 1973, *A. colemani* from France was introduced into former Czechoslovakia (Starý 1976, 2002), where it established (CABI 2007). However, no information on its impact on *M. donacis* was found.

1.33 *Melolontha melolontha* (L.), Cockchafer or May Bug (Col., Scarabaeidae)

Melolontha melolontha is native to Eurasia. It has been recorded in Europe from all countries but Albania, Cyprus, Iceland, Liechtenstein, Luxembourg, Former Yugoslav Republic of Macedonia, Norway, Portugal and Serbia (CABI 2007) and almost certainly occurs in some of these. *Melolontha melolontha* is a polyphagous species; while larvae feed on roots in the soil, adults feed on leaves, mainly of fruit trees and both broadleaved and evergreen forest trees. Major hosts of adults include *Corylus avellana* L. (hazel), *M. domestica* (apple) and *Quercus robur* L. (English oak). *Melolontha melolontha* larvae can be important pests of tree nurseries and young plantations and can cause significant damage in pastures. In outbreak years, adults feed extensively on the foliage of broadleaved and even coniferous trees.

Insecticides can be applied against *M. melolontha*, but the timing of applications is important. Despite the 3- or 4-year development cycle of the species, there may be adult flights in every year.

Natural enemies of *M. melolontha* include ten parasitiod species from nine genera, at least two predator species and eight pathogen species from seven genera. *Beauveria brongniartii* (Sacc.) is an important natural cause of mortality in *M. melolontha* populations in some countries, e.g. Switzerland and Italy (CABI 2007).

1.33.1 *Myzinum* sp. (Hym., Scoliidae)

Myzinum is a genus within the Hymenoptera family Scoliidae, a family whose larvae are ectoparasitoids on Scarabaeidae and more rarely on Curculionidae (Willemstein 1987). A *Myzinum* species has also been introduced into New Zealand to control *Costelytra zealandica* (Scarabaeidae) (BIOCAT 2005).

An unspecified *Myzinum* sp. was introduced into Poland (Lipa 1976). No information on either its origin or date of introduction is available, and no information on its establishment or impact as a biocontrol agent has been found.

1.33.2 *Tiphia* sp. (Hym., Tiphiidae)

Species in the family Tiphiidae, in particular in the genus *Tiphia*, are parasitoids of Scarabaeidae beetles and have been studied as a means to control *M. melolontha* (Györfi 1956). Worldwide, species in the genus have been introduced as biocontrol agents against 11 Scarabaeidae species (BIOCAT 2005).

An unspecified *Tiphia* sp. was introduced into Poland (Lipa 1976). No further information regarding the date of introduction, establishment or impact on *M. melolontha* was found.

1.34 *Metcalfa pruinosa* (SAY), Citrus Flatid Planthopper (Hem., Flatidae)

Metcalfa pruinosa originates from the eastern Nearctic. The species was first detected in Italy in 1979, from where it spread rapidly to adjacent countries. In 1985, it was recorded in France. *M. pruinosa* disperses over short distances by flight and over long distances through the transport of horticultural plant material. Today, the species is recorded from Croatia, France, Italy and Switzerland (CABI 2007).

Metcalfa pruinosa is a polyphagous species feeding on a wide variety of woody and herbaceous plants. In Italy, it has been recorded from over 200 different plant species (CABI 2007). Major hosts include *Citrus*, *Ficus*, *Glycine max* (L.) (soyabean), *Malus* spp., *Olea* (olive), *Prunus* (stone fruit), *P. communis* (European pear) and *Vitis* (grape) (CABI 2007). *Metcalfa pruinosa* feeds on fruits/pods and stems. Dense populations of nymphs cause stunting of the shoots, while adults produce large quantities of honeydew on which sooty mould develops. Serious damage was recorded on soyabean in northern Italy, with 30–40% crop loss in 1986 (CABI 2007). Feeding on fruits results in markings and mould and can make fruits unsaleable (CABI 2007).

Chemical control against dense nymphal populations can be applied on valuable trees, but control of sooty mould by means of fungicides is usually more useful (CABI 2007). However, chemical control of adults is difficult owing to their long life and quick recolonization of insecticide-treated areas due to their high mobility (CABI 2007). Soap solutions cause almost all of the young stages of the pest to fall to the ground (CABI 2007).

Natural enemies of *M. pruinosa* include one or several species of dryinid wasps. Moreover, nymphs are parasitized by a mite (*Leptus* sp.) and adults by an epipyropid moth, *Epipyrops barberiana* DYAR, neither of which kill their host (CABI 2007).

1.34.1 *Neodryinus typhlocybae* (ASHMEAD) (Hym., Dryinidae)

Neodryinus typhlocybae is an ectoparasitoid native to North America (Strauss 2009). The species is only recorded from *M. pruinosa* (CABI 2007) and host specificity testing in Europe so far confirms this narrow host range (Strauss 2009). In 1996, *N. typhlocybae* from the USA was introduced into Italy, where it etablished and started to reduce pest populations progressively. This suggests that *N. typhlocybae* will bring *M. pruinosa* under control in the long term (Malausa 1998).

In 1996, 43 adults (25 females, 18 males) from Italy were released in France, and in 1999, 200 cocoons were brought out at each of the 32 release sites (Malausa 1998). The species established well, but spread only slowly (i.e. only 100 m 2 years after release; Malausa 1998). Some impact on pest numbers was reported at that time. As in Italy, the effectivness of *N. typhlocybae* in France is expected to grow over the years (Malausa 1998).

Since 1996, *N. typhlocybae* has also been introduced for the biological control of *M. pruinosa* in Croatia, Slovenia and Switzerland (Lucchi *et al.* 2003). Populations of *M. pruinosa* have dropped in Switzerland, possibly due to parasitism by *N. typhlocybae* (M. Kenis, Delémont, 2014, personal communication).

1.35 *Myzus persicae* Sulzer, Green Peach Aphid (Hem., Aphididae)

Myzus persicae is probably of Asian origin, like its primary host plant (*Prunus persica* (L.) Stokes), but this species now occurs everywhere in the world except in regions with extreme temperature or humidity. In Europe, it is recorded in all countries except Albania, Estonia, Iceland, Liechtenstein, Luxembourg, Montenegro and Slovenia (CABI 2007), but probably occurs in most of these. Besides transport of infected plant material, aerial dispersal of winged forms over long distances is an effective mode of dispersal. *Myzus persicae* feeds on summer hosts from over 40 different plant families, including many economically important plants in the families Brassicaceae, Solanaceae, Poaceae, Leguminosae, Cyperaceae, Convolvulaceae, Chenopodiaceae, Asteraceae, Cucurbitaceae and Umbelliferae. The main winter host of *M. persicae* is *P. persica* (peach). In addition, *Prunus nigra* serves as a winter host in the USA, and other *Prunus* species, including *Prunus tenella* Batsch, *Prunus nana* (L.) Stokes, *Prunus serotina* Ehrh. and *Prunus americana* Marsh. may well be accepted. It is not clear, however, whether the sexual part of the life cycle is completed on species other than *P. persica* and *P. nigra* (CABI 2007).

Myzus persicae feeds on inflorescences, leaves and stems. Direct feeding damage, enhanced by the toxic effects of aphid saliva, can be of economic importance in some crops. Honeydew production may be economically important in greenhouse plants due to growth of black sooty mould. *Myzus persicae* is the most important aphid virus vector; it has been shown to transmit well over 100 plant virus diseases to plants of about 30 different families, including many major crops (CABI 2007). For instance, it is the most important vector of potato leafroll virus and therefore a major pest everywhere that potato is grown (CABI 2007).

Chemical control has had only limited or transient success, and resistance to a range of insecticides has developed worldwide. Transgenic potato plants show enhanced resistance to *M. persicae* (CABI 2007).

Natural enemies recorded on *M. persicae* include 45 parasitoid species from 17 genera (many *Aphidius* species), at least 87 predatory species and 16 pathogens from nine genera. The fungal pathogen *Lecanicillium lecanii* R. Zare and W. Gams is effective against *M. persicae* in a wide range of crops, particularly in greenhouses (CABI 2007).

1.35.1 *Hippodamia convergens* Guérin-Méneville, convergent lady beetle (Col., Coccinellidae)

Hippodamia convergens is a predatory beetle native to North America (Obrycki and Tauber 1982). It is a polyphagous species, feeding on a wide variety of insects including aphids, beetles, whiteflies and mites. In Europe, it was also introduced into Italy against unspecified aphids (Section 1.55.3). Besides Europe, *H. convergens* was further introduced into Australia, Bermuda, Chile, Hawaii, Kenya, Mexico, New, Zealand, Peru, the Philippines and South Africa against various aphids, into Ecuador against *I. purchasi* (Margarodidae) and into Chile against unspecified Coccidae (BIOCAT 2005). Today, it is recorded in Europe from Albania, Belgium, the Czech Republic, Denmark and Sweden (Roy and Migeon 2010).

In 1919, 25,000 *H. convergens* from California were introduced into France, but the species did not establish (Greathead 1976). Again in 1925, additional specimens were received to establish a rearing, and to release beetles into the field, but apparently *H. convergens* again failed to establish (Greathead 1976).

1.36 *Ophelimus maskelli* (Ashmead) (Hym., Eulophidae)

Ophelimus maskelli is a gall-forming wasp that is native to Australia and that feeds on *Eucalyptus* spp. It was first recorded in Europe in Italy in 2000 and subsequently in Greece (2002), Spain (2003), France (2005), Portugal (2006) and probably in the UK (CSL 2007; Branco *et al.* 2009). Females attack immature leaves in the lower canopy where the leaves are relatively large while still green. Each egg induces one gall and, at high population densities, the entire upper surface of leaves can be covered with galls. *Eucalyptus* trees can almost completely lose their foliage as a consequence of being attacked by *O. maskelli* (CSL 2007). No chemical control measures are described in the literature.

1.36.1 *Closterocerus chamaeleon* (Girault) (Hym., Eulophidae)

Closterocerus chamaeleon is a parasitoid native to Australia. It has a narrow host range including *O. maskelli* and perhaps some other *Ophelimus* spp. associated with *Eucalyptus* spp. (Protasov *et al.* 2007).

The first releases of *C. chamaeleon* were made in 2006 in Italy. Field studies indicated that in 2007 the rate of parasitism was >95%, and by 2008 the density of *O. maskelli* had dropped from approximately 30 galls/cm^2 leaf blade to 6–10 galls/cm^2 leaf blade (Sasso *et al.* 2008). In 2007, *C. chamaeleon* was recorded from Portugal, about 2700 km west of the release sites in Italy (Branco *et al.* 2009).

A *Closterocerus* sp. was also released in Israel; it has spread rapidly and seems to have achieved effective biological control (CSL 2007).

1.37 *Panolis flammea* (Denis and Schiffermüller), Pine Beauty (Lep., Noctuidae)

Panolis flammea is native to Eurasia. In Europe, the species is recorded from Austria, Bulgaria, Czech Republic, Estonia, Finland, Germany, Latvia, Lithuania, Poland, Slovak Republic, Sweden and the UK (CABI 2007).

Panolis flammea is restricted mainly to pine trees, and *Pinus sylvestris* L. (Scots pine) is the main host. During population outbreaks, the species may also feed on other *Pinus* species, *Picea*, *Pseudotsuga*, *Larix*, *Abies*, *Juniperus*, *Chamaecyparis*, *Quercus* and *Betula* (Sachtleben 1929; CABI 2007). Larvae feed externally on needles.

Natural enemies recorded from *P. flammea* include 16 parasitoid species from 15 genera, two congeneric predators and two pathogen species (CABI 2007).

1.37.1 *Trichogramma minutum* Riley, minute egg parasite (Hym., Trichogrammatidae)

Trichogramma minutum is an egg parasitoid native to North America (Stouthamer *et al.* 2000). The species has been recorded from 56 Lepidoptera species from different families (CABI 2007). Besides Europe, *T. minutum* was introduced as a biocontrol agent into Chile, Ecuador, Egypt, Fiji, Greece, India, Papua New Guinea, the Philiippines, New Zealand and Sri Lanka against several Lepidoptera pests (Gelechiidae, Pyralidae, Noctuidae, Oecophoridae, Tortricidae, Zygaenidae) (BIOCAT 2005). In Europe, it was also introduced against *C. pomonella* (Section 1.13.1) and *P. oleae* (Section 1.46.3) and is recorded as present in the Czech Republic, France, Germany, Greece, Italy, Spain and the UK (CABI 2007; Rasplus *et al.* 2010).

In 1933, 8 million *T. minutum* reared from specimens obtained from Barbados, Canada and the USA were released in Germany, but failed to control the pest (Greathead 1976). Greathead and Greathead (1992) mentioned an introduction of *T. minutum* into the UK, but no further information was provided.

1.38 *Parabemisia myricae* (Kuwana), Bayberry Whitefly (Hem., Aleyrodidae)

Parabemisia myricae is native to Japan. Over the past 30 years, this species has extended its geographical range dramatically, particularly in the Mediterranean region, where it spread rapidly in the 1980s. It is also recorded from countries in Africa, Central, North and South America and Oceania. In Europe, *P. myricae* occurs in Cyprus, Greece, Italy, Portugal and Spain.

Parabemisia myricae is a polyphagous species. Its major hosts include *Cinnamomum camphora* (L.) J.Presl. (camphor laurel), *Citrus*, Cucurbitaceae (cucurbits), *D. kaki* Thunb. (persimmon), *Eriobotrya japonica* (Thunb.) Lindl. (loquat), *Morus alba* L. (white mulberry) and *Pyrus* (pears). The species is also recorded from wild hosts, including *Betula* (birch), *Rhododendron* (azalea) and *Salix* (willow) species. *Parabemisia myricae* is considered to be one of the six most injurious whitefly pests, causing a problem mainly on *Citrus*. It feeds on leaves and stems, causing direct damage; moreover, sooty moulds growing on honeydew deposits block light and air from the leaves, reducing photosynthesis and productivity.

It is possible to control *P. myricae* by the fumigation of planting material, or with chemical sprays, but the latter is likely to require several successive applications because the waxy nature of the immature stages and the non-feeding period in the 'pupa' reduces their susceptibility. However, the use of pesticides is likely to make biological control, which is applied successfully in many countries worldwide, ineffective. Overall, biological control is considered the most cost-effective and sustainable method of control against this pest (CABI 2007).

Natural enemies recorded on *P. myricae* include nine parasitoid species from three genera (mainly *Encarsia* spp.) and six predatory species from five genera.

1.38.1 *Eretmocerus debachi* Rose and Rosen (Hym., Aphelinidae)

Eretmocerus debachi is presumably of Asian origin, but it was first recorded from California. *Parabemisia myricae* is the only known host (Noyes 2002). Besides Europe, the species was also introduced into Israel and Turkey against *P. myricae* (BIOCAT 2005).

In 1991, *E. debachi* was introduced into Italy, where it established (Noyes 2002) and substantial control of the pest was subsequently recorded (Barbagallo *et al.* 1992).

1.38.2 *Eretmocerus* sp. (Hym., Aphelinidae)

In addition to *E. debachi*, another *Eretmocerus* sp. was introduced into Sicily, Italy, in 1991. The species established permanently (Sinacori *et al.* 1993), but no information on its impact on *P. myricae* was found.

1.39 *Parectopa robiniella* Clemens, Locust Digitate Leafminer (Lep., Gracillariidae)

Parectopa robiniella is native to the USA. The species was introduced into Europe accidentally, with first records made in Italy in 1970 (Whitebread 1989). Today, *P. robinella* is recorded from Austria, Bosnia and Herzegovina, Croatia, Czech Republic, France, Hungary, Italy, Former Yugoslav Republic of Macedonia, Montenegro, Serbia and the Slovak Republic (CABI 2007), but is likely to occur in more countries.

Parectopa robiniella feeds exclusively on *Robinia* species. In Europe, it attacks the North American tree *Robinia pseudoacacia* (black locust), which has been planted widely throughout temperate Europe but has become invasive in some regions (Rice *et al.* 2004). Leafmining by larvae of this species causes premature leaf drop and the reduction of the photosynthetic active area (Fodor and Hâruta 2009).

Natural enemies recorded on *P. robiniella* include 11 parasitoid species from nine genera (CABI 2007). According to Fodor and Hâruta (2009), control of *P. robiniella* is accomplished by autochthonous parasitoids, entomopathogens and predators (mainly birds and spiders). High parasitism rates of over 50% in Serbia and over 60% in Italy have been reported (references in Fodor and Hâruta 2009).

1.39.1 *Closterocerus cinctipennis* Ashmead (Hym., Eulophidae)

Closterocerus cinctipennis is native to the USA (Legaspi and French 1996). Known hosts include Diptera, Hymenoptera and Lepidoptera species (Noyes 2002).

In 1972, *C. cinctipennis* was introduced into Italy, but it does not appear to have established permanently (Noyes 2002).

1.40 *Parthenolecanium corni* (Bouché), European Fruit Scale (Hem., Coccidae)

Syn.: *Lecanium corni* Bouché

The abundance of species of the *Parthenolecanium corni* complex in Europe and the apparent lack of specific parasites led to the tentative conclusion that this scale was more likely to be of North American than of European origin (Clausen 1978). Dispersal is by the first-instar crawler, aided by wind and animals, and by human transport of infested material. *Parthenolecanium corni* was also introduced into countries in South America and Oceania (CABI 2007). In Europe, the species is recorded from all countries but Cyprus, Estonia, Iceland, Ireland, Liechtenstein and Slovenia (CABI 2007).

Parthenolecanium corni is highly polyphagous, attacking some 350 plant species in 40 families. It attacks a wide range of crops, mostly woody fruit trees and ornamentals. Major hosts include *P. domestica*, *P. persica*, *Ribes nigrum* L. (blackcurrant), *Ribes rubrum* L. (redcurrant) and *Vitis vinifera* L. (grapevine). *Parthenolecanium corni* feeds mainly on leaves and stems. In the native range, the species is normally held below economic thresholds by natural enemies, but it can rapidly build up damaging populations on ornamentals, especially where natural enemies have been affected by chemical control methods. In Europe, *P. corni* can be a pest of a range of fruit and nut trees and ornamentals. Infestations result in reduced vigour and general debility of the host plant; heavy infestations may result in chlorotic spotting and premature shedding of leaves, wilting and dieback of stems. In addition, excreted honeydew forms a substrate for the growth of black sooty moulds, reducing the value of produce and ornamentals.

There are many effective insecticides causing 72–100% mortality of *P. corni*. However, in areas where effective natural enemies are active, use of chemical pesticides may upset the ecological balance and make the scale-induced damage worse (CABI 2007).

Natural enemies recorded as attacking *P. corni* include 22 parasitoid species from ten genera (several *Metaphycus* and *Microterys* spp.) and five predatory species from three genera (CABI 2007).

1.40.1 *Encyrtus fuscus* (Howard) (Hym., Encyrtidae)

Syn.: *Encyrtus californicus* (Girault)
Encyrtus fuscus is native to the USA (Jackson *et al.* 1978). Known hosts include other Coccidae, i.e. *Lecanium armeniacum*, *Lecanium hesperidum* and *Pulvinaria camellicola* (Anon. 1888/9; Howard *et al.* 1901). In Europe, it is recorded from France, Italy, Portugal, Romania, the Slovak Republic, Spain and former Yugoslavia (Noyes 2002).

In 1901, *E. fuscus* from California was introduced into Italy, but according to Clausen (1978) it failed to establish.

1.41 *Pectinophora gossypiella* (Saunders), Pink Bollworm (Lep., Gelechiidae)

Pectinophora gossypiella is native to Australia and has been introduced into many countries worldwide; it is recorded from Africa, Asia, Europe, Central, North and South America. In Europe, it is recorded as present in Albania, Bulgaria, Cyprus, Denmark, Greece, Italy, the Former Yugoslav Republic of Macedonia, Malta, Montenegro, Romania, Serbia and Spain (CABI 2007).

Pectinophora gossypiella is known from several host species of different families. Major hosts include *Abelmoschus esculentus* (L.) Moench (okra), *Abutilon indicum* (Link) Sweet (country mallow), *M. sativa* (lucerne), *Gossypium* (cotton), *Hibiscus* spp. and species in the family Malvaceae. *Pectinophora gossypiella* is a worldwide pest of cotton, and in some regions it is the key cotton pest. The species feeds on fruits/pods and inflorescences, causing failure of buds to open, fruit shedding, lint damage and seed loss. In cotton, crop losses of up to 20.2% were recorded in field trials in India during the 1970s.

Insecticidal control is hindered by the larvae being internal feeders. Genetically manipulated cotton varieties have given better yield than conventional strains in China and Mexico, but there is a risk that larvae will develop resistance.

Cultural control methods (including late planting, chemical defoliation to reduce the number of overwintering larvae, rotating crops and more) were developed to reduce the impact of the pest, mainly in cotton.

A comprehensive sterile male release programme has been conducted in Florida, USA, and infestations have since remained at low levels. Combinations of biological and chemical controls, for instance the application of *Trichogrammatoidea brasiliensis* (Ashmead) in combination with chemical insecticides, have also proved successful.

Natural enemies recorded on *P. gossypiella* include 55 parasitoid species from 22 genera, 19 predatory species from 15 genera and 16 pathogens from the genus *Bacillus* (14 of which are subspecies of *B. thuringiensis*) (CABI 2007).

1.41.1 *Chelonus* spp. (Hym., Braconidae)

Species in the genus *Chelonus* (*Microchelonus*) are known as parasitoids of Lepidoptera (see also Sections 1.41.2, 1.43.9, 2.7.1).

Between 1975 and 1976, two *Chelonus* species, one described as 'red', the other as 'black', from Ethiopia were introduced into Greece, but it is not known whether they established (Luck 1981). An additional species from Australia was introduced in 1976, but again, its establishment was not confirmed (Luck 1981).

1.41.2 *Chelonus blackburni* Cameron (Hym., Braconidae)

Chelonus blackburni is native to Hawaii (Jackson *et al.* 1978). It is an egg-larval parasitoid of Lepidoptera, particularly Microlepidoptera; at least 19 host species are recorded from 18 different genera from families including Gelechiidae, Noctuidae, Erebidae, Pyralidae and Crambidae (Oboyski 2009). Besides Europe, the species was introduced into Egypt, Mexico, Pakistan, Puerto Rico and the USA against *P. gossypiella*, into India against *Helicoverpa armigera* (Hübner) (Noctuidae) and into Fiji against *Levuana iridescens* Bethune-Baker (Zygaenidae) (BIOCAT 2005).

Between 1975 and 1976, *C. blackburni* was introduced into Greece (Luck 1981), where it has been recorded as present (CABI 2007). However, no information on its impact as a biocontrol agent was found.

1.41.3 *Pristomerus hawaiiensis* Perkins (Hym., Ichneumonidae)

Pristomerus hawaiiensis was first described from Hawaii; however, some entomologists suspect that these wasps were introduced there accidently (Oboyski 2009). It is an endoparasitoid of small caterpillars; at least 31 host species are recorded from 27 genera (including Crambidae and Tortricidae) (Oboyski 2009). Besides Europe, the species was introduced into Mexico and the USA against *P. gossypiella* (BIOCAT 2005).

Between 1975 and 1976, *P. hawaiiensis* was introduced into Greece (Luck 1981), where it has been recorded as present (CABI 2007); no information on its impact on *P. gossypiella* was found.

1.42 *Phoracantha semipunctata* (Fabricius), Eucalyptus Longhorned Borer (Col., Cerambycidae)

Phoracantha semipunctata is native to Australia. The species was first reported as an introduced insect problem in South Africa in 1906. In the later part of the 20th century, when eucalypts were widely planted as a major source of cellulose for paper production, the beetle spread worldwide into regions with a Mediterranean climate. The main cause of its rapid spread throughout the world has been the transport of eucalyptus wood. In Europe, *P. semipunctata* was first recorded in 1969 in Sardinia, Italy. Today, it is also recorded from Croatia, Cyprus, France, Malta, the Netherlands, Portugal, Spain, Sweden and Switzerland (CABI 2007). It was also introduced into countries in Africa, Asia, North and South America (CABI 2007).

Phoracantha semipunctata is restricted to *Eucalyptus* trees. Larvae mine in the stems and branches of trees. Attacked trees show some dead branches and, in some cases, the whole crown can die off. Massive beetle attacks quickly kill some *Eucalyptus* species (*E. globulus* Labill. and *E. viminalis* Labill.), while others (*E. blakelyi* Maiden) are better protected (CABI 2007). Occasionally, a single larva girdles the trunk and kills the tree. The pest causes serious damage to plantations; it is favoured by drought and by trees growing in poor soils. The average tree mortality rate between 1981 and 1983 in one province in Spain was estimated to be between 2.2 and 3.9%, equivalent to a loss of 6207 ha, despite the high economic and labour cost of control measures (CABI 2007).

Living *Eucalyptus* trees are not easily protected by pesticides, as the biology of *P. semipunctata* limits its exposure to chemicals. The adults are active during the greater part of the year and would, therefore, require either the use of pesticides with long residual effects or more frequent applications. Silvicultural measures, including the clearance and felling of all attacked and dead trees, followed by their removal, are used as the most effective and ecologically sound methods of control.

Several predators and parasitoids appear to cause significant mortality to beetles, larvae and eggs of *P. semipunctata* in Australia. Overall, seven parasitoid species from six genera, three predator and one pathogen species are recorded as natural enemies (CABI 2007).

1.42.1 *Oobius longoi* (Siscaro) (Hym., Encyrtidae)

Syn.: *Avetianella longoi* Siscaro

Oobius longoi is an egg parasitoid native to Australia, where it attacks *P. semipunctata* and the congeneric *Phoracantha recurva* Newman (Hanks *et al.* 1996); it is further recorded from *Coptocercus aberrans* Newman and *Epithora dorsalis* (MacLeay), both Cerambicidae (Noyes 2002). It has been widely distributed around the world and, besides Europe, introduced against *P. semipunctata* into South Africa, the USA and Zambia (BIOCAT 2005), as well as Chile and Uruguay (CABI 2007). It provides excellent biological control of *P. semipunctata* in some areas, but may contribute to the replacement of *P. semipunctata* by *P. recurva* in areas where both species have been introduced (CABI 2007). In Europe, it is recorded from Italy, Portugal and Spain (Noyes 2002).

In 1992, *O. longoi* was introduced into Portugal, where it established permanently (Lima and Araujo 1996); no information on the impact of the species as a biocontrol agent was found.

1.43 *Phthorimaea operculella* (Zeller), Potato Tuber Moth (Lep., Gelechiidae)

The origin of *Phthorimaea operculella* was first believed to be North America; however, Lloyd (1972) argued convincingly that the pest originated from South

America. It is a cosmopolitan pest, especially in warm temperate and tropical regions, and has been introduced into many countries in Africa, Asia, Europe and Oceania. The species was first discovered in France in 1906. Today, it is also recorded from Albania, Croatia, Cyprus, Greece, Italy, Malta, Montenegro, Portugal, Romania, Serbia, Spain and Switzerland. It has been intercepted also in the Czech Republic, Denmark, Finland, Germany, Hungary, the Netherlands, the Slovak Republic and Sweden; however, the species does not seem to be able to survive severe cold winters of temperate countries. *Phthorimaea operculella* has also been recorded in Bulgaria and the UK, but is now reported as eradicated.

Potato is the principal host for *P. operculella*, but other Solanaceae crop plants, especially tomato (*Solanum lycopersicum* L.), tobacco (*Nicotiana tabacum* L.), chilli pepper (*Capsicum frutescens* L.), aubergine (*Solanum melongena* L.), cape gooseberry (*Physalis peruviana* L.) and wild Solanaceae plants are also attacked. Females lay their eggs on foliage, soil and plant debris, or exposed tubers. Larvae mine in leaves, stems and roots (tubers). The species is a major pest in areas where the climate favours its development and where its host plants are grown on a large scale.

Many effective insecticides are known for the control of *P. operculella*, but their use may interfere with efforts to control the target species by biological means (see below). Infested seed tubers are the main cause of reinfestation; therefore, the use of healthy tubers for planting will reduce levels of field infestation. Prevention of attack is also assisted by the application of vegetable oils.

Biological control of *P. operculella* has been attempted since 1918 when *Habrobracon gelechiae* (ASHMEAD) was imported into France from the USA. Detailed analysis of data obtained through regular monitoring revealed that, in an area free from insecticides, parasitoids were a major factor in controlling *P. operculella* (Sankaran and Girling 1980). Native parasitoids can also achieve significant rates of parasitism; for example, in Sardinia, one native ichneumonid and three braconid parasitoids were recorded from *P. operculella*, with *Diadegma pulchripes* (KOKUJEV) accounting for 65.1% of the total rate of parasitism (Ortu and Floris 1989). Biological control has also been attempted successfully using nematodes (Ivanova *et al.* 1994), baculovirus (CIP 1992) and *B. thuringiensis* (Baklanova *et al.* 1990).

The list of natural enemies includes more than 60 parasitoid species, as well as numerous predators and pathogens (Rondon 2010).

1.43.1 *Agathis unicolorata* SHENEFELT (Hym., Braconidae)

Syn.: *Agathis unicolor* SCHROTTKY
Agathis unicolorata is an ectoparasitoid native to South America (Chundurwar 1977). It appears to have a very narrow host range, i.e. it is only recorded from *P. operculella* (Yu *et al.* 2005). Besides Europe, the species was also introduced into Australia, Bermuda, India, New Zealand, South Africa, St Helena, the USA, Zambia and Zimbabwe as a biocontrol agent against *P. operculella* (BIOCAT 2005).

Between 1966 and 1968, 13,295 *A. unicolorata* from India were introduced into Cyprus (Greathead 1976). The species has established in Cyprus (CABI 2007), but no information on its impact on *P. operculla* has been found.

1.43.2 *Apanteles scutellaris* Muesebeck (Hym., Braconidae)

Syn.: *Illidops scutellaris* (Muesebeck)
Apanteles scutellaris is an endoparasitoid native to North America (Sankaran and Girling 1980). Known hosts include two Gelechidae species, *Keiferia lycopersicella* (Walsingham) and *P. operculella* (Yu *et al.* 2005). Besides Europe, *A. scutellaris* was also introduced into Australia, Hawaii, India, Madagascar, New Zealand, St Helena and Zambia as a biocontrol agent against *P. operculella*, and into Hawaii and Trinidad against *K. lycopersicella* (BIOCAT 2005). In Europe, the species is recorded from Bulgaria, Cyprus and Hungary (Yu *et al.* 2005).

Between 1966 and 1968, 11,928 *A. scutellaris* from India were introduced into Cyprus (Greathead 1976). The species has established (CABI 2007), but no information on its impact on *P. operculla* has been found.

1.43.3 *Apanteles subandinus* Blanchard (Hym., Braconidae)

Apanteles subandinus is an endoparasitoid native to South America (Sankaran and Girling 1980). Known hosts include two Gelechidae species, *P. operculella* and *Tuta absoluta* Meyrick (= *Scrobipalpula absoluta* Meyr.), which is a pest of tomatoes, and a crambid, *Achyra bifidalis* (Fabricius) (Yu *et al.* 2005). Besides Europe, the species was also introduced into Australia, Bermuda, Chile, India, Madagascar, Mauritius, New Zealand, South Africa, St Helena, Tanzania, the USA, Zambia and Zimbabwe as a biological control agent against *P. operculella* (BIOCAT 2005).

Between 1966 and 1968, 17,306 *A. subandinus* from South America were introduced into Cyprus, where the species established and provided some control of *P. operculella* (Greathead 1976; Sankaran and Girling 1980).

1.43.4 *Apanteles* sp. (Hym., Braconidae)

In addition to the two *Apantheles* species mentioned above (Sections 1.43.2 and 1.43.3), an unspecified *Apanteles* sp. was introduced into Greece in 1968, but failed to establish (Greathead 1976).

1.43.5 *Habrobracon gelechiae* (Ashmead) (Hym., Braconidae)

Syn.: *Habrobracon johannseni* Viereck; *Bracon gelechiae* Ashmead; *Bracon johannseni* Viereck

Habrobracon gelechiae is native to Pakistan and presumably parts of India (Ahmad and Muzaffar 1976). This parasitoid has a broad host range, including host species from 49 genera, mainly Lepidoptera (Crambidae, Gelechiidae, Pyralidae, Noctuidae, Erebidae and Tortricidae), but it is also recorded from two Hemiptera species, *A. craccivora* KOCH (Aphididae) and *Empoasca kerri* SINGH-PRUTHI (Cicadellidae) (Yu *et al.* 2005; CABI 2007). Besides Europe, the species was also introduced into Australia, Bermuda, Hawaii, India, New Zealand, South Africa, St Helena, Zambia and Zimbabwe as a biological control agent against *P. operculella*, into Mexico and the USA against *P. gossypiella*, and into the USA against *A. transitella* (WALKER) (Pyralidae) (BIOCAT 2005).

In 1918, *H. gelechiae* was introduced from the USA into France (Greathead 1976). Further shipments were made in 1920 and 1921 to establish rearings and to make additional releases. The species established (Yu *et al.* 2005) and some impact on pest number were reported (Sankaran and Girling 1980).

In 1938, *H. gelechiae* was introduced from the USA into Malta, but failed to establish there (Greathead 1976).

Between 1965 and 1968, 30,205 *H. gelechiae* were introduced from the USA into Cyprus (CIBC 1970). According to Sankaran and Girling (1980), the species established permanently, but it was not reported to be present in Cyprus by Yu *et al.* (2005). No information on the impact of *H. gelechiae* on *P. operculella* was found.

1.43.6 *Campoplex haywardi* BLANCHARD (Hym., Ichneumonidae)

Campoplex haywardi is an endoparasitoid native to South America (BIOCAT 2005). Known hosts include two Gelechiidae species, *P. operculella* and *T. absoluta* (Yu *et al.* 2005). Besides Europe, *C. haywardi* was also introduced into Australia, Bermuda, India, Madagascar, Mauritius, New Zealand, South Africa, St Helena, Tanzania, the USA, Zambia and Zimbabwe as a biological control agent against *P. operculella* (BIOCAT 2005).

In 1965, 15,061 *C. haywardi* from South America were introduced into Cyprus, where the species established (Greathead 1976). No information on its impact on *P. operculella* was found.

1.43.7 *Copidosoma koehleri* BLANCHARD (Hym., Encyrtidae)

Syn.: *Copidosoma uruguayensis* TACHIKAWA

Copidosoma koehleri is a polyembryonic endoparasitoid native to South America (Keasara and Sadeh 2007; Segoli *et al.* 2009). Known hosts include three Gelechiidae species (*P. operculella*, *T. absoluta* and *Symmetrischema tangolias* (GYEN)) and one Tortricidae, *Grapholita molesta* (BUSCK) (Noyes 2002; CABI 2007). Besides Europe, *C. koehleri* was also introduced into Australia, Bermuda, Hawaii, India, Japan, Kenya, Madagascar, Mauritius, New Zealand, the Seychelles, South Africa,

St Helena, Tanzania, the USA, Venezuela, Yemen, Zambia and Zimbabwe as a biological control agent against *P. operculella* (BIOCAT 2005). According to Rasplus *et al.* (2010), in Europe the species is recorded from Albania, Cyprus, Greece and Italy.

In 1947, *C. koehleri* was introduced into Italy, but it failed to establish (Bartoloni 1952). In 1994, specimens from South Africa were shipped to Italy, where they were mass reared to perform inundative releases. However, results from experimental studies with *C. koehleri* showed only low rates of parasitism of *P. operculella* larvae under open-field conditions (maximum 3%; Pucci *et al.* 2003).

In 1965, a small number of *C. koehleri* was first released in Cyprus. In the following year, 874,500 specimens were sent from India to Cyprus and released (CIBC 1970; Greathead 1976). It is uncertain whether the species has established permanently in Cyprus (CABI 2007).

In 1968, *C. koehleri* from South Africa was introduced into Greece. According to Greathead (1976), it failed to establish, but Rasplus *et al.* (2010) recorded it as present in Greece.

1.43.8 *Eriborus trochanteratus* (Morley) (Hym., Ichneumonidae)

Eriborus trochanteratus is native to India and Sri Lanka (Perera 1977). Known hosts include four Lepidoptera species, i.e. *Corcyra cephalonica* (Stainton) (Pyralidae), *Nephantis serinopa* Meyrick, *Opisina arenosella* Walker (Oecophoridae) and *P. operculella* (Perera 1977; Yu *et al.* 2005; CABI 2007). Besides Europe, the species was also introduced into New Zealand, St Helena, Tanzania and Zambia as a biological control agent against *P. operculella*, and into India against *O. arenosella* (Sankaran and Girling 1980; BIOCAT 2005). It has also been used to make augmentative releases against *O. arenosella* in Sri Lanka (Cock and Perera 1987). In 1966, *E. trochanteratus* was introduced from India into Cyprus, where it established (Sankaran and Girling 1980). However, no information on its impact on *P. operculella* was found.

1.43.9 *Chelonus curvimaculatus* Cameron (Hym., Braconidae)

Syn.: *Microchelonus curvimaculatus* (Cameron)
Chelonus curvimaculatus is an endoparasitoid native to Africa (Kfir 1997). It is considered polyphagous; known hosts include at least 26 Lepidoptera species from 23 genera from several different families (Yu *et al.* 2005). Besides Europe, the species was also introduced into Bermuda, New Zealand, St Helena and the USA as a biological control agent against *P. operculella*, and into the USA against *P. gossypiella* (Sankaran and Girling 1980; BIOCAT 2005).

In 1965, *C. curvimaculatus* from India and South Africa was introduced into Cyprus, where it established (Sankaran and Girling 1980). No information on its impact on *P. operculella* was found.

1.43.10 *Orgilus lepidus* Muesebeck (Hym., Braconidae)

Orgilus lepidus is an endoparasitoid native to South America (Sankaran and Girling 1980). According to Yu *et al.* (2005), *O. lepidus* has only been reared from *P. operculella*. Outside of Europe, the species was also introduced into Australia, Bermuda, India, New Zealand, South Africa, St Helena, Tanzania, the USA and Zambia for the biological control of *P. operculella*, and into the USA against *P. gossypiella* (Sankaran and Girling 1980; BIOCAT 2005).

In 1965, *O. lepidus* from South America was introduced into Cyprus. It established here, but no information on its impact was found (Sankaran and Girling 1980).

1.43.11 *Orgilus parcus* Turner (Hym., Braconidae)

Orgilus parcus is an endoparasitoid native to South Africa (Sankaran and Girling 1980). Known hosts include *P. operculella* and *Spodoptera exigua* (Hübner) (Noctuidae) (Yu *et al.* 2005). Besides Europe, *O. lepidus* was also introduced into Bermuda, India, New Zealand, St Helena, the USA and Zambia as a biological control agent of *P. operculella* (BIOCAT 2005; Yu *et al.* 2005).

In 1967/68, 40,425 specimens of *O. lepidus* were introduced from India into Cyprus, where it established (Greathead 1976). No information on the impact of this species on *P. operculella* was found.

1.43.12 *Temelucha* sp. (Hym., Ichneumonidae)

Species in the genus *Temelucha* are larval parasitoids. Outside of Europe, this unidentified *Temelucha* sp. was also introduced into Australia, Bermuda, India, St Helena, the USA and Zambia for the biological control of *P. operculella* (Sankaran and Girling 1980; BIOCAT 2005). It is assumed that the different releases were made with the same species.

In 1965, *Temelucha* sp. from South America was introduced into Cyprus, where it established (Sankaran and Girling 1980), but no information on its impact on *P. operculella* was found.

1.43.13 *Trichogramma* spp. (Hym., Trichogrammatidae)

Trichogramma species are egg parasitoids frequently used as biocontrol agents. In 1944, unidentified *Trichogramma* species, which were imported from the USA into Spain for the control of *C. pomonella* (Section 1.13.1), were also released for the biological control of *P. operculella* (Urquijo 1945). Some 18,000 parasitized *S. cerealella* eggs were brought out in potato fields. The use of *Trichogramma* species is considered an important factor in the decline of *P. operculella* in north-western Spain (Urquijo 1951).

In 1966, an unspecified *Trichogramma* sp. was introduced into Greece, but it failed to establish (BIOCAT 2005). According to Greathead (1976), three more unspecified *Trichogramma* spp. were released in 1970.

1.44 *Phyllocnistis citrella* Stainton, Citrus Leafminer (Lep., Gracillariidae)

Phyllocnistis citrella is native to Asia and was first discovered outside its native range in 1993 in Florida, USA. One year later, it was detected in the Mediterranean basin. Today, *P. citrella* is found in all of the world's citrus-growing areas. In Europe, it is recorded from Croatia, Cyprus, France, Greece, Italy, Malta, Montenegro, Portugal, Serbia and Spain.

Phyllocnistis citrella is a major pest of *Citrus*, but also attacks other Rutaceae. Larvae mine in leaves, and heavy infestations can hinder the growth of newly planted trees or reduce fruit production of mature trees. *Phyllocnistis citrella* also appears to faciliate the spread of *Citrus* bacterial canker, *Xanthomonas axonopodis* pv. *citri* (Hasse), since leaves damaged by the larvae show an increased susceptibility to infection by this disease.

Many insecticides are effective against *P. citrella*, but they are also liable to reduce the numbers of natural enemies that may keep the pest populations below acceptable thresholds.

A large number of parasitoids are known to attack *P. citrella*, including 18 parasitoid species from 12 different genera and two predatory species.

1.44.1 *Ageniaspis citricola* Logvinovskaya (Hym., Encyrtidae)

Ageniaspis citricola is an endoparasitoid native to Thailand (Longo and Siscaro 1997). It appears to have a very narrow host range, since it has so far only been recorded from *P. citrella* (Longo and Siscaro 1997; Noyes 2002). Besides Europe, *A. citricola* was also introduced into Algeria, Australia, Bahamas, Brazil, Honduras, Israel, Japan, Morocco, Oman, Peru, Syria, Tunisia, Turkey, Venezuela and the USA as a biological control agent against *P. citrella* (BIOCAT 2005). Most introductions were successful, although *A. citricola* does not perform well in the arid Mediterranean climate. In Europe, the species is reported from Cyprus, France, Italy, Poland and Spain (Noyes 2002; Rasplus *et al.* 2010).

In the 1990s, *A. citricola* was introduced into Cyprus, where it established permanently according to Schauff *et al.* (1998); no information on its impact on *P. citrella* was found.

Between 1996 and 1997, *A. citricola* was introduced into Sicily, Italy, where the species has established; no information on its impact as a biocontrol agent was found. In 1996, some 1000 *A. citricola* were also introduced from Florida into mainland Italy, but no evidence of its establishment there was found (Longo and Siscaro 1997).

Between 1996 and 1997, *A. citricola* was introduced from Florida, USA, into France, but it is not known if it has established there (Longo and Siscaro 1997).

Between 1996 and 1999, *A. citricola* was introduced into Spain. According to Vercher *et al.* (2000, 2003), the species failed to establish, but it has been recorded as present according to other sources (CABI 2007; Rasplus *et al.* 2010).

In 1996, *A. citricola* from Cyprus was introduced into Greece, but no evidence of its establishment was found (Longo and Siscaro 1997).

1.44.2 *Cirrospilus ingenuus* Gahan (Hym., Eulophidae)

Syn.: *Cirrospilus quadristriatus* (Subba Rao and Ramamani)
Cirrospilus ingenuus is an ectoparasitoid native to South-east Asia (China, Thailand) (Schauff *et al.* 1998). Known hosts include *P. citrella* and *Rhynchaenus mangiferae* Fabricius, a curculionid leafminer (Coleoptera) (Noyes 2002). Laboratory observations indicate that *C. ingenuus* can attack citrus leafminer larvae that have been parasitized by *A. citricola* and hence act as a facultative hyperparasitoid, but it is not clear whether this also occurs under field conditions (Hoy and Nguyen 1997). Besides Europe, *C. ingenuus* was also introduced in Oman, Morocco, Syria, Tunisia and the USA as a biological control agent against *P. citrella* (BIOCAT 2005). In Europe, the species is recorded as present in Cyprus, Spain and Portugal (Rasplus *et al.* 2010).

In the 1990s, *C. ingenuus* was introduced into Cyprus, where it established permanently according to Schauff *et al.* (1998), but not according to Noyes (2002). No information on the impact of *C. ingenuus* on *P. citrella* was found.

Between 1996 and 1999, *C. ingenuus* was also introduced into Spain (Vercher *et al.* 2000, 2003), but it apparently failed to establish (Noyes 2002).

1.44.3 *Citrostichus phyllocnistoides* (Narayanan) (Hym., Eulophidae)

Citrostichus phyllocnistoides is an ectoparasitoid native to Asia (Afghanistan, India, Oman, Pakistan, southern China) (Wang *et al.* 2006). Known hosts include *P. citrella* and *Trioza obsoleta* (Buckton) (Psyllidae, Hemiptera) (Noyes 2002). Besides Europe, *C. phyllocnistoides* was also introduced into Australia and Israel as a biological control agent against *P. citrella*, but failed to establish (Schauff *et al.* 1998; BIOCAT 2005). In Europe, the species is recorded from Bulgaria, France, Germany, Greece, Italy, the Netherlands, Poland, Portugal, Romania, Spain, Switzerland and the former USSR (CABI 2007; Rasplus *et al.* 2010).

In the 1990s, *C. phyllocnistoides* was introduced into Cyprus, where it established permanently (Schauff *et al.* 1998); no information on the impact of the species was found.

Between 1996 and 1999, *C. phyllocnistoides* was also introduced into Spain, where it established and dispersed, reaching high parasitism rates and lowering citrus leafminer population densities (Garcia-Mari *et al.* 2003). In 1996, *C. phyllocnistoides* from Cyprus was introduced into Greece, but no evidence of its establishment was found (Michelakis 1997).

1.44.4 *Galeopsomyia fausta* La Salle (Hym., Eulophidae)

Galeopsomyia fausta is an ectoparasitoid native to the Neotropics (Llácer *et al.* 2005). It recruited *P. citrella* as a host in many South American countries invaded by *P. citrella* (Argentina, Brazil, Colombia, Honduras, Mexico, Nicaragua and Puerto Rico) (La Salle and Peña 1997). No information was found on the original hosts of *G. fausta.* In Central and South America, the activity of this species has resulted in good biological control of *P. citrella* (Llácer *et al.* 2005).

Between 1996 and 1999, *G. fausta* was also introduced into Spain, but the species failed to establish (Vercher *et al.* 2000, 2003).

1.44.5 *Quadrastichus citrella* Reina and La Salle (Hym., Eulophidae)

Quadrastichus citrella is an ectoparasitoid native to Asia. It attacks second- and third-instar larvae of *P. citrella*, which is the only host so far recognized (Reina and La Salle 2004).

In the 1990s, *Q. citrella* from Thailand was introduced into Cyprus, where it established permanently (Schauff *et al.* 1998); no information on the impact of the species on *P. citrella* was found.

Between 1996 and 1999, *Q. citrella* was also introduced into Spain, where it has been recorded as permanently established (Vercher *et al.* 2000; Llácer *et al.* 2006).

Between 1996 and 1997, *Q. citrella* was introduced into Sicily, Italy, but failed to establish (Siscaro *et al.* 2003). In 1996, 3000 *Q. citrella* from Israel were also introduced into mainland Italy (Longo and Siscaro 1997). The species established and seems to have a good adaptability to the climatic condidions of southern Italy (Longo and Siscaro 1997). No information on its impact as a biocontrol agent was found.

1.44.6 *Semielacher petiolatus* (Girault) (Hym., Eulophidae)

Syn.: *Pseudiglyphella petiolata* Girault

Semielacher petiolatus is an ectoparasitoid native to Australia (Michelakis 1997). So far, it has only been recorded from *P. citrella.* Besides Europe, *S. petiolatus* was also introduced into Algeria, Israel, Morocco, Oman, Syria, Tunisia and Turkey as a biological control agent against *P. citrella* (BIOCAT 2005). In Europe, the species is recorded from Cyprus, Greece, Italy, Portugal and Spain (Noyes 2002; Rasplus *et al.* 2010).

In the 1990s, *S. petiolatus* was introduced into Cyprus, where it established permanently (Schauff *et al.* 1998); no information on the impact of the species on *P. citrella* was found.

In 1996, *S. petiolatus* was introduced from Cyprus into Greece. According to Michelakis (1997), there is no evidence of its establishment in Greece, but it has been reported as present by Noyes (2002).

Between 1996 and 1997, *S. petiolatus* was introduced accidentally, along with other parasitoids of *P. citrella*, into Sicily, Italy, where it became established (Siscaro *et al.* 1999, 2003); no information on the impact of the species was found.

Between 1996 and 1999, *S. petiolatus* was also introduced into Spain (Vercher *et al.* 2000, 2003), but again no information on its impact on *P. citrella* was found.

1.45 *Planococcus citri* (Risso), Citrus Mealybug (Hem., Pseudococcidae)

Planococcus citri is native to the South Pacific, Australia and South-east Asia (Cox 1981), but today it has an almost worldwide distribution. In Europe, it is recorded from Austria, Belgium, Bulgaria, Cyprus, the Czech Republic, Estonia, Latvia, Lithuania, France, Germany, Greece, Hungary, Italy, Malta, the Netherlands, Poland, Portugal, Romania, the Slovak Republic, Spain and the UK (CABI 2007). In southern Europe, northern America and southern Australia, it occurs mainly in greenhouses (CABI 2007).

Planococcus citri is polyphagous and occurs on a wide range of flowering plants. Major host plants include *Citrus* and *Ziziphus mauritiana* Lam. (jujube). The species is reported as a serious pest on various crops, including *Citrus*, grapes and mangoes. It is the most injurious of the mealybugs on *Citrus* in the Mediterranean region. In temperate regions, it occurs mainly on greenhouse plants, but also occurs outdoors in the summer. *Planococcus citri* attacks fruits/pods, inflorescence, leaves, roots and stems. Feeding leads to general wilting of the plants due to sap depletion. Furthermore, infestation also causes indirect physical damage because of the sooty moulds that develop on the honeydew secreted by the mealybug.

In several countries, *P. citri* has been controlled successfully with chemicals and natural enemies, or a combination of the two. Attempts have also been made to control the pest using semiochemicals, cultural methods and resistant plant material.

Natural enemies recorded to attack *P. citri* include 25 parasitoid species from 15 genera, 32 predator species from 22 genera and two pathogen species (CABI 2007).

1.45.1 *Coccidoxenoides peregrinus* (Timberlake) (Hym., Encyrtidae)

Syn.: *Pauridea peregrina* Timberlake
Coccidoxenoides peregrinus is a solitary endoparasitoid originally known from south China (Berlinger 1977). Known hosts include at least ten other Pseudococcidae

species (Noyes 2002). Besides Europe, the species was also introduced into Bermuda, Chile and the USA for the biological control of *P. citri* and into Ghana against *Planococcoides njalensis* (LAING) (BIOCAT 2005). It is further recorded as introduced in Kenya and Peru. In Europe, the species is reported from Cyprus and Italy (Noyes 2002).

Between 1956 and 1957, *C. peregrinus* from China was released in Italy. According to Zinna (1960b), it failed to establish, but Noyes (2002) reported it as established.

1.45.2 *Coccophagus gurneyi* COMPERE (Hym., Aphelinidae)

Coccophagus gurneyi is an endoparasitoid native to Australia (Parkes and Walter 2001). Known hosts include 11 other Pseudococcidae species and *A. aurantii* (Diaspididae), but it is reported to hyperparasitize two encyrtid wasps, *Leptomastix dactylopii* HOWARD and *Tetracnemoidea peregrina* (COMPERE) (Noyes 2002). Besides Europe, *C. gurneyi* was also introduced into Ghana for the biological control of *P. njalensis*, into Chile, South Africa, the USA and the former USSR against *Pseudococcus calceolariae*, into the former USSR against *Pseudococcus comstocki*, as well as into the Cook Islands, Hawaii and New Zealand against unspecified Pseudococcidae (BIOCAT 2005).

Coccophagus gurneyi from California have been introduced into Italy (no year of introduction is provided), but it is not known whether the species ever established (Domenichini 1959).

1.45.3 *Cryptolaemus montrouzieri* MULSANT, mealybug ladybird (Col., Coccinellidae)

Cryptolaemus montrouzieri is a predatory beetle native to eastern Australia. It is a polyphagous species with both larvae and adults feeding on a wide range of Coccoidea prey, but also Aphididae species (CABI 2007). *Cryptolaemus montrouzieri* has been mass-produced for biological control for over 100 years. Imported from Australia in 1891 and used first for the control of mealybugs on *Citrus* in California, it has since been introduced thoughout all continents and has become established wherever climates are suitable. In Europe, it is recorded as present in Albania, Greece, Italy, Portugal, Russia and Sweden (CABI 2007; Roy and Migeon 2010). Its establishment in France is considered uncertain (CABI 2007).

In 1908, *C. montrouzieri* from California was first released in Italy, but failed to establish (Costantino 1935). Between 1919 and 1920, a further 200 adults from France were released. The species established, but was not abundant in 1927. In 1935, 1200 adults from Spain were released both on the mainland and on the islands of Sicily and Sardinia. Altogether, *C. montrouzieri* contributed to reduce the levels of *P. citri*, but appeared to require augmentation to achieve control. Winter mortality and perhaps also predation by ants limited its population development in Italy (Greathead 1976).

In 1919, *C. montrouzieri* from the USA was introduced into France, but establishment was described as difficult because of low winter temperatures (Malausa *et al.* 2008).

In 1927, *C. montrouzieri* from France was introduced into Spain, where a rearing was established and individuals repeatedly released into the field. By 1951, some 1,055,131 colonies had been distributed in Spain (Greathead 1976). Releases of ten adults per tree, coupled with ant control, proved effective against *P. citri* when released as soon as the pest was noticed (Gómez Clemente 1954b). However, as in Italy, *C. montrouzieri* did not persist in the field because the host disappeared from the branches in the winter. Furthermore, extremes of temperature and humidity are chiefly responsible for both heavy mortality of adults and inhibiting reproduction (Ruiz Castro 1965).

In 1929, 3960 *C. montrouzieri* from Spain were introduced into Portugal, from where some, but not complete, control of the pest was reported (Ferreira 1939).

In 1933, 900 *C. montrouzieri* from Spain were introduced into Greece. On Corfu, *P. citri* disappeared as a pest after the release of *C. montrouzieri* (Greathead 1976). Again, in 1964/65, *C. montrouzieri* from Spain and in 1969 from France were introduced into Greece to establish mass rearing. While Argyriou (1970) reported permanent establishment of *C. montrouzieri* in Spain, Pelekassis (1974) and CABI (2007) considered it as not permanently established.

In 1954, 800 *C. montrouzieri* from California were introduced into Cyprus (Greathead 1976). The majority of the beetles were released, but some were kept to start a rearing. Another shipment from California followed in 1956 (Greathead 1976). As no recoveries were made in the field, further introductions of *C. montrouzieri* were made from Italy (Sicily) in 1966 and from India in 1968. All releases resulted only in temporary establishment, because the beetles did not survive the winter. Overall, the use of *C. montrouzieri* to control *P. citri* on Cyprus was considered unsuccessful (Greathead 1976).

1.45.4 *Leptomastix dactylopii* Howard (Hym., Encyrtidae)

Leptomastix dactylopii is an endoparasitoid native to South America, probably Brazil (CABI 2007). Its host range includes at least 33 Coccoidea species (Pseudococcidae and Coccidae) (Noyes 2002). It was first introduced as a biological control agent into California, USA, in 1934. The species has since been spread around the world. Besides Europe, *L. dactylopii* was also introduced into Australia, Chile, India, Israel, South Africa, the USA and the former USSR against *P. citri*, into Ghana against *P. njalensis* and into Hawaii against *Dysmicoccus brevipes* (BIOCAT 2005). In Europe, it is recorded as present in Austria, Belgium, Bosnia and Herzegovina, Cyprus, the Czech Republic, Denmark, Finland, France, Germany, Greece, Ireland, Italy, the Netherlands, Norway, Poland, Portugal, Spain, Sweden, the UK and former Yugoslavia (CABI 2007; Rasplus *et al.* 2010).

In 1948, *L. dactylopii* was introduced from California into Spain, where a rearing was established (Greathead 1976). The species established and was considered the most important parasitoid on *P. citri* in southern Spain. Together with

Anagyrus bohemani (Westwood), a native parasitoid, it achieved substantial control of *P. citri* (Ruiz Castro 1965). Field studies revealed, however, that the ant *Lasius niger* (Latreille) could reduce parasitism of *P. citri* by 35% (Campos *et al.* 2006). According to Cavalloro and di Martino (1986), another introduction of *L. dactylopii* was made in 1977, but no further information was found.

In 1953, *L. dactylopii* was introduced into France and 100,000 adults released (Greathead 1976). The species was again introduced in 1972 (Malausa *et al.* 2008), where it established and, in combination with the predator *C. montrouzieri*, achieved successful control of *P. citri* (CABI 2007).

In 1956, *L. dactylopii* was introduced into Italy to establish a rearing and to release it in the field (Zinna 1960a,b). At some release sites, the species established, while at other sites it reproduced successfully but did not survive the winter (Russo 1958; Zinna 1960a,b). An additional release was made in Sicily (no date provided), which resulted in parasitism rates between 60 and 95% and effective control of *P. citri* (Cavalloro and di Martino 1986). In general, successful control of *P. citri* is reported in cases where a combination of *L. dactylopii* and the predator *C. montrouzieri* (Section 1.45.3) has been released (CABI 2007).

In 1967, *L. dactylopii* was introduced into Cyprus, where a rearing was established and subsequently 30,000 adults released (Greathead 1976). The species did not survive the winter, and therefore an additional introduction of individuals from Italy was made in 1977. Rate of parasitism by *L. dactylopii* increased from 9% a few months after its release to 15% in 1979 (Krambias and Kontzonis 1980).

1.45.5 *Nephus reunioni* Fürsch (Col., Coccinellidae)

Syn.: *Scymnus reunioni* Fürsch

Nephus reunioni is a predatory beetle native to South Africa (Cavalloro and di Martino 1986). It is recorded from three Pseudococcidae and one Diaspididae (*A. aurantii*; CABI 2007). *Nephus reunioni* was also introduced into Israel and the former USSR for the biological control of *P. citri* and *Nipaecoccus viridis* (Newstead), into France for the control of *Pseudococcus viburni* (Section 1.48.1), and into the former USSR against Pseudococcidae species in general (BIOCAT 2005). In Europe, it is recorded from Albania, France, Greece, Italy, Portugal and Spain (CABI 2007; Roy and Migeon 2010).

In 1982, specimens of *N. reunioni* originating from South Africa were introduced into Sicily, Italy (Cavalloro and di Martino 1986), and later on specimens originating from La Réunion into Sardinia (Ortu and Prota 1986). *Nephus reunioni* failed to establish in Sicily, while it established permanently in Sardinia (BIOCAT 2005). No information on the impact of *N. reunioni* on *P. citri* was found.

1.45.6 *Tetracnemoidea brevicornis* (Girault) (Hym., Encyrtidae)

Syn.: *Tetracnemus pretiosus* Timberlake; *Hungariella pretiosus* (Timberlake)

In 1906, lilac twigs bearing parasitized *P. pentagona* from the USA were introduced into Italy and *E. berlesei* was reared from these (Clausen 1978). About 30 females were subsequently released on infested mulberry trees. Additional shipments were made from the USA until 1908 (Clausen 1978). In addition, infested mulberry twigs were shipped from Japan to Italy in 1908 and attached directly to branches of infested trees in the field. By 1910, *E. berlesei* was firmly established and had already spread over 2 km from the initial release site (Clausen 1978).

In 1910, *E. berlesei* was collected in Italy and introduced into Switzerland and Austria, and in 1914 it was also introduced into Spain (Clausen 1978). *Encarsia berlesei* soon became established throughout the *P. pentagona*-infested regions of these countries (Clausen 1978). The introduction of *E. berlesei* into Europe resulted in complete control of *P. pentagona* in most regions, and by 1916 the pest had become rare and far below the threshold of economic injury (Clausen 1978).

In 1918, *E. berlesei* from Italy was released in France, where it kept *P. pentagona* under control in Mediterranean areas. However, in colder regions (Lyon), *E. berlesei* emerged too late to control the pest completely (Delucchi 1976). Resurgence of *P. pentagona* has been reported since 1948, presumably because of the adverse effects of insecticide treatments on *E. berlesei* (Clausen 1978).

1.47.3 *Pteroptrix orientalis* (Silvestri) (Hym., Aphelinidae)

Syn.: *Archenomus orientalis* Silvestri

Pteroptrix orientalis is an endoparasitoid native to the Far East. It is also recorded from *C. dictyospermi* (Morgan) (Noyes 2002).

In 1909 (EPPO 2014), or between 1924 and 1925 (Greathead 1976), *P. orientalis* originating from Japan (EPPO 2014) or the Far East (Greathead 1976) was introduced into Italy, from where substantial impact on *P. pentagona* was reported (Greathead 1976). In an unmanaged mulberry orchard in a coastal region sampled in the early 1990s, *P. orientalis* (Silvestri) was the third most abundant parasitoid, after *A. proclia* and *E. berlesei* (Sections 1.47.1 and 1.47.2; Pedata *et al.* 1995).

1.47.4 *Rhyzobius lophanthae* (Blaisdell), scale-eating ladybird (Col., Coccinellidae)

Syn.: *Lindorus lophanthae* (Blaisdell)

Rhyzobius lophanthae is a predatory beetle native to Australasia (Clausen 1978; Roy and Migeon 2010). Its host range includes 23 scale insect species (mainly Diaspididae, but also Coccidae and Pseudococcidae) (CABI 2007). This beetle was also released into Bermuda, Chile, the Cook Islands, India, Kenya, Marianas (Guam), Mauritius (Agelega Island), Peru, the Philippines, St Vincent, South Africa, Tanzania, Trinidad, Uganda, the USA and the former USSR against other scale insects (BIOCAT 2005). In Europe, *R. lophanthae* was also introduced into former Czechoslovakia against *D. perniciosus* (Section 1.18.3). It is reported today from Albania, Cyprus, France, Germany, Greece, Italy, Malta, Portugal, Spain and the UK (BIOCAT 2005; Roy and Migeon 2010). Its presence in France is considered to be a result of natural dispersal (Malausa *et al.* 2008).

In 1908, *R. lophanthae* was introduced into Italy (Greathead 1971). The species established, but no information on its impact on *P. pentagona* was found.

1.48 *Pseudococcus viburni* (Signoret), Obscure Mealybug (Hem., Pseudococcidae)

Syn.: *Pseudococcus affinis* Maskell
Pseudococcus viburni is believed to originate from South America (Charles 2011), but today it is a cosmopolitan pest occurring in different regions in Africa, Asia, Europe and Oceania (CABI 2007). In Europe, the species is recorded as present in France, Italy, Spain and the former USSR, but these records are not confirmed (CABI 2007).

Pseudococcus viburni is a polyphagous species feeding on the phloem of vines and woody-stemmed plants; it is an important pest on agricultural and horticultural plants (*Citrus*, apple, grapevine, stone fruits) and ornamentals, as well as on tubers of *Dahlia* sp. and potatoes (Karamaouna and Copland 2000). *Pseudococcus viburni* is also a vector for infectious pathogens.

Synthetic pesticides are used to control *P. viburni*. While the waxy coating and the species' habit of seeking sheltered feeding sites protect it from water-based pesticides, oil-based organophosphate pesticides are effective in reducing its population density. However, some *P. viburni* populations in New Zealand have developed resistance to organophosphate pesticides (Hamlet 2005).

Natural enemies recorded on *P. viburni* include 12 parasitoid species from eight genera and seven predatory species from five genera (CABI 2007).

1.48.1 *Nephus reunioni* Fürsch (Col., Coccinellidae)

Syn.: *Scymnus reunioni* Fürsch
Nephus reunioni is a predatory beetle native to South Africa (Cavalloro and di Martino 1986). The species is recorded from three other Pseudococcidae and one Diaspididae (*A. aurantii*; CABI 2007). *Nephus reunioni* was also introduced into Italy (Section 1.45.5), Israel and the former USSR for the biological control of *P. citri* and *N. viridis*, as well as against Pseudococcidae in general (BIOCAT 2005). In Europe, it is recorded from Albania, France, Greece, Italy, Portugal and Spain (CABI 2007; Roy and Migeon 2010).

In 1974, *N. reunioni* originating from La Réunion was introduced into mainland France, where the species established temporarily (OPIE 1986). It was also sent to other countries (OPIE 1986), but it remains unclear whether it was released elsewhere in Europe. No information on its impact on *P. viburni* was found.

1.49 *Rhyacionia buoliana* (Denis and Schiffermüller), European Pine Shoot Moth (Lep., Tortricidae)

Rhyacionia buoliana is native to Europe and has been introduced to other continents accidentally. Today, it is recorded in Asia, Africa, and North and South America (CABI 2007). In Europe, it is reported from all countries but Albania, Croatia, Iceland, Liechtenstein, Luxembourg and Malta (CABI 2007).

Rhyacionia buoliana attacks species of the genus *Pinus* with two or three needles per fascicle (Diploxylon group). *Pinus radiata* D. Don (radiata pine) is reported as its major host, but it also attacks many other *Pinus* species and *Pinus menziesii* (Mirb.) Franco (Douglas fir). Larvae feed internally in leaves and stems. The damage caused by *R. buoliana* does not kill trees; however, infestation of the apices inflicts important damage, including a loss in height and diameter, and stem deformations. Plantations aged less than 15 years are considered to be the most susceptible to *R. buoliana*. In western Europe, *R. buoliana* is considered to be a sporadic but serious pest, with an outbreak observed at the beginning of the 20th century (CABI 2007).

Several insecticides can be used against *R. buoliana* (Butcher and Haynes 1960).

There are many species of natural enemies associated with *R. buoliana*, including 41 parasitoid species from 27 genera and two predatory species (CABI 2007).

1.49.1 *Hyssopus thymus* Girault (Hym., Eulophidae)

Hyssopus thymus is an ectoparasitoid native to North America. The known host range includes several Tortricidae, but also *Dioryctria auranticella* (Grote) (Pyralidae) and *Metzneria lappella* (L.) (Gelechiidae) (Noyes 2002). *Hyssopus thymus* adopted *R. buoliana* as a host after its introduction into North America (Greathead 1976).

Between 1965 and 1969, *H. thymus* originating from Canada and the USA were released in Germany (Greathead 1976). Further releases were made in Lower Saxony in 1970 (1800 individuals) and in 1972 (900 individuals). In Baden Württemberg, 650 individuals were released in 1970 and 500 in 1973 (Bogenschütz 1972; Gelmroth 1972; Greathead 1976). The species is reported as established (CABI 2007), but no information on its impact on *R. buoliana* has been found.

1.49.2 *Itoplectis conquisitor* (Say) (Hym., Ichneumonidae)

Itoplectis conquisitor is a pupal parasitoid native to North America (CABI 2007). It is a polyphagous species; it is reported to parasitize at least 157 species from 104 genera (Yu *et al.* 2005), including various Lepidoptera but also *Cassida rubiginosa* (Chrysomelidae; Coleoptera). *Itoplectis conquisitor* is also known to hyperparasitize *Bathyplectes curculionis* (Ichneumonidae; Hymenoptera) (CABI 2007). The species adopted *R. buoliana* as a host after its introduction into North America (Greathead

1976). *Itoplectis conquisitor* was also released in 1970 against unspecified Lepidoptera pests in Poland (Section 1.57.1).

Between 1965 and 1969, *I. conquisitor* originating from Canada and the USA were released in Germany. Further releases were made in Lower Saxony in 1970 (10,300 individuals) and in 1971 (2780 individuals), and in Baden Württemberg in 1970 (200 individuals; Bogenschütz 1972; Biermann 1973). The species is reported as established in Germany (CABI 2007), but no information on its impact on *R. buoliana* has been found.

1.50 *Saissetia oleae* (Olivier), Black Olive Scale (Hem., Coccidae)

Saissetia oleae is native to Africa. All life stages of this insect can be transported easily on ornamental plants or propagation material and it is today recorded from countries in Asia, Europe, Central, North and South America and Oceania. In Europe, it is recorded from Austria, Bulgaria, Cyprus, Denmark, France, Germany, Greece, Italy, Malta, the Netherlands, Portugal, Spain, Switzerland and the UK (CABI 2007).

Saissetia oleae is polyphagous and has been recorded from 113 host-plant species in 49 families. Major hosts include *Asparagus officinalis* L. (asparagus), *Citrus*, *Coffea arabica* (arabica coffee), *Erica* (heaths), *Gossypium* (cotton), *Ilex* (holly), *Nerium* (oleander), *Olea* (olive), *Prunus* (stone fruit), *Pyrus* (pears), *Rosmarinus officinalis* (rosemary) and *Tamarix* (tamarisk). The species feeds on leaves and stems. Removal of large quantities of sap debilitates the plant and can cause wilting, desiccation of tissues and dieback. Sooty mould growth on honeydew deposits blocks light and air from the leaves and impairs photosynthesis. Such damage reduces the overall yield and quality of the produce. *Saissetia oleae* is one of the most important pests of *Citrus* in the Mediterranean Basin (CABI 2007). It has been suggested that there are several biologically different races of *S. oleae* (Bartlett 1960).

According to Copland and Ibrahim (1985), the irregular feeding and waxy dorsal surface of *S. oleae* make it difficult to control with pesticides.

Natural enemies recorded on *S. oleae* include at least 65 parasitoid species from 24 genera (many *Coccophagus* species), six predatory species and one pathogen species (*L. lecanii* R. Zare and W. Gams) (CABI 2007). It is thought that the different physiological races of *S. oleae* are attacked by different natural enemies (Bartlett 1960). In many countries, the release and establishment of exotic parasitoids mainly from southern Africa resulted in a considerable degree of control.

1.50.1 *Coccophagus ceroplastae* (Howard) (Hym., Aphelinidae)

Syn.: *Aneristus ceroplastae* Howard

Coccophagus ceroplastae is an endoparasitoid native to eastern Asia (Japan, Pakistan, Taiwan; Waterhouse 1998; Waterhouse and Sands 2001; BIOCAT 2005). Known hosts include at least 29 other Coccidae species, a Diaspididae (*Aspidiotus destructor* Signoret) and two Pseudococcidae species (*Pseudococcus adonidum* Westwood and *Nipaecoccus nipae* (Maskell)) (Noyes 2002). Besides Europe, *C. ceroplastae* was also introduced into Bermuda, Japan and the USA as a biological control agent in *Citrus*, guava and olive plantations (BIOCAT 2005).

Between 1973 and 1974, *C. ceroplastae* originating from Pakistan was released in France (OPIE 1986). The species apparently failed to establish as it is not now recorded from France (Noyes 2002).

Coccophagus ceroplastae was also introduced into Crete and Greece from France, but no date of introduction was provided (Paraskakis *et al.* 1980). The species established but largely failed to control *S. oleae*, probably because of insecticide use (Paraskakis *et al.* 1980).

1.50.2 *Coccophagus rusti* Compere (Hym., Aphelinidae)

Coccophagus rusti is native to Africa (Kenya, Uganda; CABI 2007). Besides *S. oleae*, known hosts include at least 28 other Coccidae species, a Diaspididae (*A. destructor*) and two Pseudococcidae species (*N. nipae* and *P. adonidum*) (Noyes 2002). *Coccophagus rusti* was introduced into Europe, Israel and the USA for the biological control of *S. oleae* in *Citrus* and olive plantations, and into Guam against *Saissetia coffeae* (Walker) and *Parasaissetia nigra* (Nietner) (BIOCAT 2005).

In 1968, about 200 *C. rusti* from California were introduced into Greece, but no recoveries were subsequently made (Greathead 1976).

1.50.3 *Diversinervus elegans* Silvestri (Hym., Encyrtidae)

Diversinervus elegans is a solitary endoparasitoid native to Africa (Paraskakis *et al.* 1980; Waterhouse and Sands 2001). Known hosts include at least 18 Coccidae species (Noyes 2002). Besides Europe, *D. elegans* was also introduced into Australia and the USA as a biological control agent in *Citrus* and olive plantations (BIOCAT 2005).

Between 1973 and 1979, about 1 million adult *D. elegans* from Israel were released in seven areas in Crete, Greece (Paraskakis *et al.* 1980). Limited recoveries were subsequently made at six of the seven release sites, but the species was reported to have only a low impact on pest numbers. Its effectiveness was probably reduced by insecticides (Paraskakis *et al.* 1980).

Between 1972 and 1975, *D. elegans* was also released in France, where it established but only reached low densities (OPIE 1986; Malausa *et al.* 2008).

1.50.4 *Encyrtus aurantii* (Geoffroy) (Hym., Encyrtidae)

Syn.: *Encyrtus lecaniorum* (Mayr)
Encyrtus aurantii is an endoparasitoid native to Pakistan and India (BIOCAT 2005; Malausa *et al.* 2008). Known hosts include the target pest and other Coccidae

(*Coccus hesperidum* L. and *Nippaecoccus viridis*) (CABI 2007). Besides Europe, *E. aurantii* was also introduced into the USA as a biological control agent against *C. hesperidum* in *Citrus* (BIOCAT 2005). In Europe, the species is recorded as present in Austria, Cyprus, the Czech Republic, France, Greece, Italy, Spain, the former USSR and former Yugoslavia (Noyes 2002; CABI 2007).

In 1953, *E. aurantii* was released in France, together with two other parasitoids, *Moranila californica* and *Scutellista caerulea* (Sections 1.50.11, 1.50.13; Greathead 1976). The species was again introduced in 1973–1975 (Malausa *et al.* 2008). *Encyrtus aurantii* established, but its impact as a biocontrol agent was unknown (Greathead 1976).

Individuals originating from France were also released in Greece. While *E. aurantii* reproduced on *S. oleae* on potatoes in the laboratory, it largely failed to attack the pest in the field. The low impact was suspected to be at least partly due to insecticides (Paraskakis *et al.* 1980).

1.50.5 *Metaphycus anneckei* Guerrieri and Noyes and *Metaphycus hageni* Daane and Caltagirone (Hym., Encyrtidae)

Metaphycus anneckei and *Metaphycus hageni* are endoparasitoids native to southern Africa (Guerrieri and Noyes 2000). The two species have been confused with each other in the past and European introductions of *M. anneckei* were misidentified as *Metaphycus lounsburyi* (Guerrieri and Noyes 2000). Information on which species was introduced and where is not completely reliable and, therefore, both species are dealt with together here. While *S. oleae* is the only host recorded for *M. hageni*, the host range of *M. anneckei* includes other Coccidae, i.e. *C. hesperidum*, *Saissetia miranda* (Cockerell and Parrott) and *Waxiella mimosa* (Signoret) (Guerrieri and Noyes 2000). Besides Europe, *M. hageni* was also introduced into the USA against *S. oleae* (BIOCAT 2005), and *M. anneckei* into Australia, Iran, Israel, New Zealand, several countries in South America and the USA against the same target pest (BIOCAT 2005). *Metaphycus anneckei* was further introduced into Guam against *S. coffeae* and *P. nigra* (BIOCAT 2005). In Europe, *M. hageni* is recorded from Denmark, France, Greece, Italy, Portugal and Spain, and *M. anneckei* from Cyprus, Greece, Italy, Poland, Portugal and Spain (Guerrieri and Noyes 2000; Rasplus *et al.* 2010).

In 1971, *M. anneckei* was released in Italy (Guerrieri and Noyes 2000). No information on the impact of the species on *S. oleae* was found.

Apparently, one or both species were also released in Portugal (Guerrieri and Noyes 2000), but neither the date of introduction nor information on their impact on *S. oleae* was found.

Metaphycus anneckei was introduced into Greece accidentally (Parastakis *et al.* 1980) and complete control of the pest has been reported (Cavalloro and di Martino 1986).

According to OPIE (1986), *M. anneckeii* was also released and established in France in 1980; however, the species was not listed in Malausa *et al.* (2008).

1.50.6 *Metaphycus helvolus* (Compere) (Hym., Encyrtidae)

Metaphycus helvolus is an endoparasitoid native to southern Africa. The species has been confused with both *M. hageni* and *M. anneckei* in the past (Guerrieri and Noyes 2000). Information on its introduction into individual countries might therefore not be completely reliable. *Metaphycus helvolus* is a parasitoid of a large number of species of soft scales, including *C. aonidum* (Diaspididae) and other Coccidae (*C. hesperidum*, *Coccus pseudomagnoliarum* (Kuwana), *Coccus viridis*, *P. nigra*, *Protopulvinaria pyriformis* (Cockerell) and *S. coffeae*) (CABI 2007). Besides Europe, *M. helvolus* was also introduced into Australia, Chile, Ecuador, Iran, Israel and the former USSR as a biological control agent against *S. oleae* (BIOCAT 2005). The species was further introduced into mainland USA against *C. pseudomagnoliarum*, into Hawaii and Guam against *P. nigra* and into Guam against *S. coffeae* (BIOCAT 2005). In Europe, *M. helvolus* is recorded from Austria, Belgium, Cyprus, France, Greece, Italy, the Netherlands, Portugal, Spain, Sweden, Switzerland, the UK and the former USSR (CABI 2007).

In 1962, *M. helvolus* originating from California was released on Crete, Greece. By 1967 it was found in all samples up to 24 km away from the release site, apparently replacing *M. flavus* (Greathead 1976). Further releases were made on other islands (Prastakis *et al.* 1980; Cavalloro and di Martino 1986). The species established and spread on two of the three islands, and substantial or complete control of the pest species is reported from there.

Metaphycus helvolus originating from California was released in mainland France in 1969 and on Corsica in 1971 (OPIE 1986; Guerrieri and Noyes 2000). The introductions were successful and some impact on pest numbers has been recorded. *Metaphycus helvolus* also attacks *C. hesperidum*, *C. pseudomagnoliarum*, *P. pyriformis* and *Protopulvinaria mesembryanthemi* (Vallot) in the field in France (Malausa *et al.* 2008).

In 1970, *M. helvolus* originating from California was first introduced into Sicily (Delucchi 1976) and 1 year later to mainland Italy (Viggiani 1978). The species established, but a few years after its release it was reported that no control had yet resulted (Delucchi 1976). A few years later, Longo (1984) found that in Sicily *M. helvolus* was among the most common parasitoids of *S. oleae*, and more recently Delrio and Foxi (2010) reported that in Sardinian olive groves *M. helvolus* was well established and provided, together with *M. lounsburyi* (Howard) (Section 1.50.7), biological control of *S. oleae*.

In 1976, *M. helvolus* was released in Spain, from where substantial control of pest numbers is reported (Cavalloro and di Martino 1986, Tena *et al.* 2008).

Between 1988 and 1989, *M. helvolus* was introduced from France into Cyprus and mass reared; subsequently, 2800 individuals were released (Orphanides 1993). Due to the combined effect of *M. helvolus* and *M. lounsburyi* (Section 1.50.7), *S. oleae* population densities have reduced from outbreak levels to almost non-existence and have remained at these low levels (Orphanides 1993).

1.50.7 *Metaphycus lounsburyi* (Howard) (Hym., Encyrtidae)

Syn.: *Metaphycus bartletti* Annecke and Mynhardt
Metaphycus lounsburyi is an endoparasitoid native to South Africa (Bellows *et al.* 1999). The species has been confused with both *M. hageni* and *M. anneckei* in the past, and releases of this species in Europe were under the name *Metaphycus bartletti* (Guerrieri and Noyes 2000). *Metaphycus lounsburyi* is also recorded as a natural enemy of other Coccidae, i.e. *Ceroplastes*, *Coccus*, *Filippia* and other *Saissetia* species (Noyes 2002). Besides Europe, *M. lounsburyi* was also introduced into Australia, Israel and the USA as a biological control agent against *S. oleae* (BIOCAT 2005). This species is considered among the most successful introductions against *S. oleae* worldwide (Clausen 1978). In Europe, *M. lounsburyi* is recorded from Albania, Cyprus, Denmark, France, Greece, Italy, the Netherlands, Poland and Spain (Noyes 2002; Rasplus *et al.* 2010).

In 1973, *M. lounsburyi* was introduced into mainland France, and in 1980 into Corsica. The species established and some impact on pest numbers was recorded (OPIE 1986).

In 1979, *M. lounsburyi* was introduced into Crete, Greece, but it did not establish according to Paraskakis *et al.* (1980); however, it has been recorded as present in Greece according to CABI (2007).

In 1979, the species was also introduced into Italy, where it established permanently (Viggiani and Mazzone 1983); no information on its impact on *S. oleae* was found. Also in 1979, several hundred individuals from Israel were released in Spain (Guerrieri and Noyes 2000; Tena *et al.* 2008), where it established permanently, but no information on its impact was found.

Between 1988 and 1989, *M. lounsburyi* from France was introduced into Cyprus (Orphanides 1993). A total of 81,000 individuals were released (Orphanides 1993). The species turned out to be very efficient; due to the combined effect of *M. lounsburyi* and *M. helvolus* (Section 1.50.6), *S. oleae* population densities reduced from outbreak level to almost non-existence and remained at low levels after the discontinuation of these releases (Orphanides 1993).

1.50.8 *Metaphycus stanleyi* Compere (Hym., Encyrtidae)

Metaphycus stanleyi is a gregarious endoparasitoid native to South Africa (Clausen 1978). Its host range includes various Coccidae, i.e. *Ceroplastes*, *Coccus*, *Eucalymnatus*, *Filippia*, *Parasaissetia*, *Pulvinaria* and other *Saissetia* species (Noyes 2002). Besides Europe, *M. stanleyi* was also introduced into Hawaii, Peru and the USA as a biological control agent against *S. oleae* (BIOCAT 2005). It was further introduced into Bermuda, Canada, Chile, Guam and Israel against other Coccidae (BIOCAT 2005). In Europe, the species is recorded from Greece, Italy and Spain (Noyes 2002).

In 1960, *M. stanleyi* from California was introduced into Italy, where it did not establish according to Clausen (1978), but was considered to be present according to CABI (2007).

1.50.9 *Metaphycus swirskii* Annecke and Mynhardt (Hym., Encyrtidae)

Metaphycus swirskii is an endoparasitoid native to Africa. Its host range includes *S. oleae* and nine other Coccidae species (Noyes 2002). Besides Europe, *M. swirskii* was also introduced into Israel for the biological control of *S. oleae* and other Coccidae (BIOCAT 2005). In Europe, the species is recorded from France, Greece, Italy, the Netherlands and Spain (Noyes 2002; CABI 2007; Rasplus *et al.* 2010).

Between 1976 and 1979, a total of 600,000 *M. swirskii* from Israel were introduced into Crete, Greece (Paraskakis *et al.* 1980). The species was recovered only occasionally and had little impact on pest numbers, probably because of insecticides (Paraskakis *et al.* 1980).

In 1977, *M. swirskii* was introduced into Italy, where the species established and from where some impact on pest numbers has been recorded (OPIE 1986).

Between 1977 and 1979, *M. swirskii* from South Africa was also introduced into France, including Corsica, and became permanently established (OPIE 1986; Malausa *et al.* 2008); no information on its impact on *S. oleae* was found.

1.50.10 *Microterys nietneri* (Motschulsky) (Hym., Encyrtidae)

Syn.: *Microterys flavus* (Howard)

Microterys nietneri is an endoparasitoid native to Africa (Waterhouse and Sands 2001). Recorded host species include at least 25 Coccidae species of the genera *Eucalymnatus*, *Parasaissetia*, *Parthenolecanium*, *Pulvinaria* and *Saissetia* (Noyes 2002). Besides Europe, *M. nietneri* was also introduced into Australia, New Zealand and the USA against *S. oleae* (BIOCAT 2005). It was further released in Bermuda against *Pulvinaria psidii* Maskell, in Australia and the former USSR against *C. hesperidum* and in Japan against *Ceroplastes rubens* Maskell (BIOCAT 2005). In Europe, the species is reported from Bulgaria, Cyprus, France, Germany, Hungary, Italy, Portugal, Spain and former Yugoslavia (Noyes 2002; Rasplus *et al.* 2010).

In 1960, *M. nietneri* from California was introduced into Italy, where it failed to establish according to Clausen (1978). The species is, however, recorded as present in Italy by Noyes (2002).

1.50.11 *Moranila californica* (Howard) (Hym., Pteromalidae)

Moranila californica is an egg predator and an external parasitoid when eggs are not available. The host range of this North American species also includes other Coccidae, i.e. *Ceroplastes floridensis* Comstock and *C. rubens* (Baker and Hardy 2005; CABI 2007). Besides Europe, *M. californica* was also introduced into Australia and the USA as a biological control agent against *S. oleae* (BIOCAT 2005). It was further released in Australia and Japan against *C. rubens* (CABI 2007). In Europe, the species is recorded from France, Greece, Italy and Spain (CABI 2007; Rasplus *et al.* 2010).

In 1953, *M. californica* from California was released in France, but no information on either its establishment or its impact on *S. oleae* has been reported (Greathead 1976).

1.50.12 *Rhyzobius forestieri* (Mulsant) (Col., Coccinellidae)

Rhyzobius forestieri is a predatory beetle native to Australia (Katsoyannos 1984). Host species include several other Coccidae (*C. hesperidum*, *C. viridis*, *Milviscutulus mangiferae* (Green) and *Parthenolecanium persicae* (Fabricius)) (CABI 2007). Besides Europe, *R. forestieri* was introduced into Israel and New Zealand against *S. oleae* (BIOCAT 2005). It was further introduced into the Cape Verde Islands against *C. viridis* and into Hawaii against an unspecified Pseudococcidae (BIOCAT 2005). In Europe, the species is recorded from Albania, Cyprus, France, Greece and Italy (CABI 2007; Roy and Migeon 2010).

Between 1980 and 1986, *R. forestieri* from California was introduced and released in France (Iperti *et al.* 1989). Up to 50 adults per olive tree were released and the species established and provided substantial control of *S. oleae* (Iperti *et al.* 1989). *Rhyzobius forestieri* arrived naturally in France from Italy, but was also introduced in 1986 from Crete, Greece (Malausa *et al.* 2008).

First pilot releases of *R. forestieri* in Greece were made in 1981 with specimens from California (Katsoyannos 1984). Between 1982 and 1983, the species was released at 22 localities and studies were conducted on the efficiency of *R. forestieri* as a biological control agent. The species established and played a key role in bringing the *S. oleae* infestations under control; by April 1983, the density of *S. oleae* in the olive grove had been reduced about 100 times (Katsoyannos 1984).

Rhyzobius forestieri was also released in Cyprus, but there was no information on its establishment or impact on *S. oleae* (Cyprus 1985).

1.50.13 *Scutellista caerulea* (Fonscolombe) (Hym., Pteromalidae)

Syn.: *Scutellista cyanea* Motschulsky

Scutellista caerulea is an egg predator native to Africa (Paraskakis *et al.* 1980). Recorded host species also include 12 other Coccidae, mainly *Ceroplastes* spp. (CABI 2007). Besides Europe, *S. caerulea* was also introduced into Australia, Chile, New Zealand, Peru, the USA and the former USSR as a biological control agent against *S. oleae* (BIOCAT 2005). It was further released in Australia, Guam, Hawaii, Israel, Japan, the USA and the former USSR against other Coccidae (BIOCAT 2005). In Europe, the species is recorded from Cyprus, France, Greece, Italy, Spain and the former USSR (CABI 2007).

In 1953, *S. caerulea* was released in France, but no information on its impact on *S. oleae* was reported (Greathead 1976).

Although an indigenous strain of *S. cyanea* was already present in Crete, Greece , 2500 adults from Israel were released in 1977/78 (Paraskakis *et al.*1980).

Since the two strains could not be distinguished, it was not known whether the new strain became established, but even if it did, it apparently had very little impact (Paraskakis *et al.*1980).

1.51 *Stictocephala bisonia* Kopp and Yonke, Buffalo Treehopper (Hem., Membracidae)

Syn.: *Ceresa bubalus* (Fabricius); *Ceresa alta* Walker

Stictocephala bisonia is native to North America (Greathead 1976). It became established in Italy in the late 1940s, from where it spread to other countries, probably by infested nursery stock (CABI 2007). Today, *S. bisonia* is recorded from Europe in Albania, Austria, Bosnia and Herzegovina, Bulgaria, Croatia, the Czech Republic, France, Greece, Hungary, Italy, Former Yugoslav Republic of Macedonia, Montenegro, Portugal, Romania, Serbia, the Slovak Republic, Spain and Switzerland (CABI 2007) and more recently in Poland (Świerczewski and Stroiński 2011).

Stictocephala bisonia is a polyphagous species feeding on at least 42 genera of herbaceous plants belonging to 13 families, and 29 genera of woody plants belonging to an additional 16 families (CABI 2007). Major host plants include *Malus* (apple) and *Prunus* (stone fruit) species, *P. communis* (European pear) and *V. vinifera* (grapevine). The preferred nymphal host is lucerne. *Stictocephala bisonia* feeds on leaves and stems. Fruit trees are most susceptible to attack because damaged twigs affect fruit production. Young trees can be killed if attacked heavily for several successive years. Grapevines develop corky growths and *S. bisonia* causes severe dieback. Damage is most acute where legumes are grown as an understorey crop in orchards or vineyards.

To control this pest, nurseries and young orchards should be weeded and undercropping with members of the vetch family (Fabaceae) under trees should be avoided (CABI 2007).

The only natural enemy for which information could be found in the literature was the egg parasitoid *Polynema striaticorne.*

1.51.1 *Polynema striaticorne* Girault (Hym., Mymaridae)

Polynema striaticorne is an egg parasitoid native to North America (Greathead 1976). Its host range also includes the nabid *Nabis americoferus* (CABI 2007). Besides Italy, the species is recorded as also present in Georgia (CABI 2007).

In 1965, *P. striaticorne* was collected in the USA and a rearing culture established in Italy (Vidano and Meotto 1968). Field releases started in April 1966 and continued in successive years. For later releases, an improved mass-rearing technique was used to obtain large numbers for release so as to accelerate the spread of the parasitoid (Greathead 1976). The species has established permanently and high rates of parasitism have been achieved in some places after continued releases (up to a mean parasitism level of 73.4%) (CABI 2007). According to Vidano *et al.*

(1987), *S. bisonia* disappeared slowly but almost completely from the vineyards of Piedmont and other Italian regions where *P. striaticorne* was spread directly or arrived by means of young fruit trees.

1.52 *Toxoptera aurantii* (Boyer de Fonscolombe), Black Citrus Aphid (Hem., Aphididae)

Toxoptera aurantii is native to the oriental region (BIOCAT 2005). Today, the species is an important economic pest, widely distributed in tropical and subtropical regions. In Europe, it is recorded from Cyprus, France, Greece, Italy, Malta, Portugal and Spain (CABI 2007). It was further introduced in various countries in Africa, Central, North and South America and Oceania.

Toxoptera aurantii is a highly polyphagous species recorded from more than 120 plant species, mainly in the families Anacardiaceae, Annonaceae, Araliaceae, Euphorbiaceae, Lauraceae, Malvaceae, Moraceae, Rubiaceae, Rutaceae and Theaceae. It is particularly damaging to the tender new vegetation on *Citrus* trees (Katsoyannos *et al.* 1997). Other hosts of economic importance include cocoa, fig, mango, *Thea*, *Annona*, *Camellia*, *Gardenia*, *Cydonia japonica*, *Cinchona* and *Magnolia*. The species infests young leaves and flowers, thereby often affecting plant growth as well as the yield and quality of fruit and tea when infestations are serious (CABI 2007). *Toxoptera aurantii* is also an important vector of the tristeza virus of citrus.

There are several effective pesticides against *T. aurantii*. In general, if chemical control is to be used, it should be applied as soon as the pest occurs.

Natural enemies recorded on *T. aurantii* include 18 parasitoid species from ten genera, 26 predators and two pathogen species (genus *Entomophthora*) (CABI 2007).

1.52.1 *Harmonia axyridis* Pallas, Asian ladybird (Col., Coccinellidae)

Harmonia axyridis is a polyphagous predator native to central and eastern Asia. It has been used widely as a biological control agent of pest aphids and scale insects (see also Section 1.5.1).

In 1993, a colony of *H. axyridis* was imported from France into Greece, and in 1994 insectary-reared adults were released in 11 *Citrus* orchards in four areas of Greece (Katsoyannos *et al.*, 1997). In total, 620 adults were released, 30–40/tree. In 1994, *H. axyridis* was recovered from seven localities (Katsoyannos *et al.* 1997), but according to Brown *et al.* (2008) it probably established only temporarily.

1.52.2 *Lysiphlebus testaceipes* (Cresson) (Hym., Braconidae)

Lysiphlebus testaceipes is a solitary endoparasitoid of aphids and native to North America. In Europe, the species was also released against other aphid pests, including *A. spiraecola* and *A. spiraephaga* (Sections 1.5.2, 1.6.2, 1.55.4). *Lysiphlebus testaceipes* is today recorded from Albania, Bosnia and Herzegovina, Bulgaria, Croatia, Czech Republic, Denmark, France, Greece, Italy, Former Yugoslav Republic of Macedonia, Montenegro, Portugal, Serbia and Spain (Yu *et al.* 2005; Rasplus *et al.* 2010).

Lysiphlebus testaceipes was imported from former Czechoslovakia (original material from Cuba; Section 1.6.2) to Mediterranean France in 1973/74 (Cavalloro and di Martino 1986; Malausa *et al.* 2008), from where substantial control of *T. aurantii* has been reported (BIOCAT 2005). It was also released and became established on Corsica (OPIE 1986).

In 1982, *L. testaceipes* was introduced to Italy and became established (Ortu and Prota 1986), but no information on its impact on *T. aurantii* was found.

The species was also introduced against *T. aurantii* into Spain (no date of introduction given), from where some impact on pest numbers was reported (OPIE 1986).

1.53 *Trialeurodes vaporariorum* (Westwood), Greenhouse Whitefly (Hem., Aleyrodidae)

Trialeurodes vaporariorum is a cosmopolitan species (Raworth *et al.* 2002). It is a widely distributed pest of ornamental and horticultural plants, mainly in greenhouses, and occurs in Africa, North, Central and South America and Oceania (CABI 2007). In Europe, it is recorded from Austria, Belgium, Bosnia and Herzegovina, Bulgaria, Croatia, the Czech Republic, Denmark, Estonia, France, Germany, Greece, Hungary, Italy, Former Yugoslav Republic of Macedonia, Malta, Montenegro, the Netherlands, Poland, Portugal, Romania, Serbia, Spain, Sweden, Switzerland and the UK (CABI 2007).

The total number of recorded *T. vaporariorum* host plants is at least 859 species, belonging to 469 genera in 121 families. In temperate countries, the most severely affected crops are aubergine, cucumber, beans, sweet peppers, tomatoes and a large number of ornamentals, including species of *Fuchsia*, *Gerbera*, *Pelargonium*, *Solanum*, chrysanthemums, poinsettias and primulas. *Trialeurodes vaporariorum* feeds on fruits/pods, inflorescence, leaves and stems. It damages plants directly by sucking sap from leaves and indirectly by transmitting viruses and producing a sticky honeydew, which prevents crops from functioning normally and acts as a substrate for fungal growth (sooty moulds). *Trialeurodes vaporariorum* became an economically important pest insect of greenhouse vegetable and ornamental crops in the mid-1970s in Beijing, China. More recently, it has become a serious horticultural pest within areas of southern Europe and may be responsible for the increase in the incidence of tomato chlorosis virus.

Many recommended insecticides give some control, but strains of *T. vaporariorum* have developed resistance to many of them. Studies have further demonstrated the effectiveness of UV absorbent plastic films in reducing *T. vaporariorum* infestations on protected crops (Diaz *et al.* 2006).

Trialeurodes vaporariorum is attacked by parasitoid species of the genera *Encarsia* and *Eretmocerus* and by fungal pathogens, many of which have been used as biological control agents in glasshouses. Lacewings (*Chrysoperla* spp.) are also used as predators of whiteflies in greenhouses. In total, natural enemies recorded on *T. vaporariorum* include 17 parasitoid species from four genera, 24 predatory species from 16 genera and ten pathogen species from five genera (CABI 2007).

1.53.1 *Encarsia pergandiella* Howard (Hym., Aphelinidae)

Encarsia pergandiella is a endoparasitoid native to the new world (Greenberg *et al.* 2008). The species is also recorded from 14 other Aleyrodidae species and is reported to hyperparasitize Aphelinidae (Hymenoptera) species, including *E. formosa*, *Encarsia inaron* (Walker), *Encarsia lutea* (Masi), *Encarsia meritoria* Gahan and *Encarsia tricolor* (Noyes 2002). *Encarsia pergandiella* was also used in Isreal for the biological control of *T. vaporariorum* (BIOCAT 2005). In Europe, the species is reported from France and Italy (Noyes 2002).

Between 1979 and 1984, *E. pergandiella* was released in Italy and became established (Mazzone and Viggiani 1985), but no information on its impact on *T. vaporariorum* was found.

1.54 *Unaspis yanonensis* (Kuwana), Arrowhead Scale (Hem., Diaspididae)

Unaspis yanonensis originates from South-east Asia. It was introduced accidentally into limited areas in southern France (Provence, Alpes, Côte d'Azur) around the 1960s and into Italy around 1986. Other European countries in which this species has been recorded include Switzerland and the UK (CABI 2007). It is further recorded from Australia and Fiji.

Unaspis yanonensis attacks only *Citrus* spp. The species feeds on fruits, inflorescence, leaves and stems. Attacked fruits lose their commercial value because of the feeding punctures. Attacks on branches and leaves lead to leaf fall and possibly to complete dieback. The cause of the dieback is not yet understood, but it has been suggested that the scale may inject toxic saliva into the plant tissue. In Europe, *U. yanonensis* was rare in commercial citrus-growing areas, but it has recently been recorded in the two main citrus cultivation areas of Calabria (Campolo *et al.* 2013). Overall, only a small spread of *U. yanonensis* has been observed in past 25 years in the Mediterranean region, so it has recently been removed from the list of the EPPO A2 quarantine pests.

Chemical control (organophosphorus compounds, oil, insect growth regulators) is applied against *U. yanonensis*, but effective control is difficult.

Unaspis yanonensis is attacked by numerous parasitoids, mycoparasites and predators. Natural enemies recorded include seven parasitoids from three genera (mainly *Aphytis* species), ten predatory species from five genera and two pathogen species (CABI 2007).

1.54.1 *Aphytis yanonensis* DeBach and Rosen (Hym., Aphelinidae)

Aphytis yanonensis is an ectoparasitoid native to China (Furuhashi and Nishino 1983). The species is also recorded from two other Diaspididae, i.e. *Unaspis citri* (Comstock) and *Unaspis euonymi* (Comstock) (CABI 2007). It was also released in Argentina and Japan against *U. yanonensis*, and in the USA against two other Diaspididae, *Parlatoria pergandii* Comstock and *U. citri* (BIOCAT 2005).

In 1984, individuals from China were released in France (OPIE 1986), where, in combination with *Coccobius fulvus* (Section 1.54.2), satisfactory results were obtained (CABI 2007). However, it has been noted that although *A. yanonensis* persists, in some situations it has to be augmented (Audant *et al.* 2005).

According to Rasplus *et al.* (2010), *A. yanonensis* is also reported from Greece, but no information on its release in this country has been found.

1.54.2 *Coccobius fulvus* (Compere and Annecke) (Hym., Aphelinidae)

Coccobius fulvus is an endoparasitoid native to China (CABI 2007). It is also recorded from six other Diaspididae and from *P. citri* (Pseudococcidae) (Noyes 2002). The species was also released in Japan against *U. yanonensis* and in the USA against another Diaspididae, *Aulacaspis yasumatsui* Takagi (BIOCAT 2005).

In 1984, *C. fulvus* from Japan was introduced into France (OPIE 1986; Malausa *et al.* 2008). In combination with *A. yanonensis* (Section 1.54.1), satisfactory results were obtained (CABI 2007).

1.55 Unspecified Aphids (Hem., Aphidae)

1.55.1 *Coccinella* sp. (Col., Coccinellidae)

Coccinella spp. are predatory beetles, feeding mainly on aphids. In 1908, one or more '*Coccinella* spp.' from Hawaii was introduced into Italy (Greathead 1976), but it has been assumed that none became established.

1.55.2 *Harmonia axyridis* Pallas, Asian ladybird (Col., Coccinellidae)

Harmonia axyridis is a polyphagous predator native to central and eastern Asia. It has been used widely as a biological control agent of pest aphids and scale insects (see also Sections 1.5.1 and 1.52.1).

In 1982, *H. axyridis* from China was introduced into France, but the introduced genotype appeared to have difficulty becoming established in southern France (Malausa *et al.* 2008). From 2000 onwards, another genotype was reported to have colonized northern parts of France (Malausa *et al.* 2008).

1.55.3 *Hippodamia convergens* Guérin-Méneville, convergent lady beetle (Col., Coccinellidae)

Hippodamia convergens is a predatory beetle native to North America (Obrycki and Tauber 1982). It is a polyphagous species feeding on a wide variety of insects including aphids, beetles, whiteflies and mites (CABI 2007). In Europe, it was also introduced into France against *M. persicae* (Section 1.35.1). Besides Europe, *H. convergens* was further introduced into Australia, Bermuda, Chile, Hawaii, Kenya, Mexico, New Zealand, Peru, the Philippines and South Africa against aphids, into Ecuador against *I. purchasi* (Margarodidae) and into Chile against unspecified Coccidae (BIOCAT 2005). Today, it is recorded in Europe from Albania, Belgium, the Czech Republic, Denmark and Sweden (Roy and Migeon 2010).

In 1908, *H. convergens* from Hawaii and California were introduced into Italy, but failed to establish there (Greathead 1976).

1.55.4 *Lysiphlebus testaceipes* (Cresson) (Hym., Braconidae)

Lysiphlebus testaceipes is a solitary endoparasitoid of aphids and is native to North America. In Europe, the species was also released against other aphid pests, i.e. against *A. spiraecola*, *A. spiraephaga* and *T. aurantii* (Sections 1.5.2, 1.6.2, 1.52.2).

Lysiphlebus testaceipes was introduced into Portugal against unspecified aphids (Yu *et al.* 2005), but no further information on this release was available.

1.56 Unspecified Scale Insects (Hem., Coccidae)

1.56.1 *Aphytis chrysomphali* (Mercet) (Hym., Aphelinidae)

Aphytis chrysomphali is an ectoparasitoid native to the Far East (Greathead 1976; Orphanides 1984). It is one of the most polyphagous and widespread *Aphytis* species known, having been reared from more than 50 Diaspididae and Coccidae hosts (Noyes 2002). The majority of records concern *A. aurantii*, *C. dictyospermi* and *A. destructor*. Besides Europe, *A. chrysomphali* was also introduced into Australia, Bermuda, Chile, Fiji, Hawaii, Israel, South Africa and the former USSR against six Diaspididae pests (BIOCAT 2005). In Europe, the species is recorded from Belgium, Cyprus, France, Greece, Hungary, Italy, Spain and former Yugoslavia (Noyes 2002; CABI 2007).

Between 1924 and 1925, *A. chrysomphali* from the Far East was introduced into Italy (Greathead 1976), but no information on its impact as a biocontrol agent was found. No information was found on whether the species was released deliberately in other Mediterranean countries or if it dispersed naturally.

1.56.2 *Aphytis lingnanensis* Compere (Hym., Aphelinidae)

Aphytis lingnanensis is a gregarious ectoparasitoid native to China, India and Pakistan (CABI 2007). The species was also introduced as a biological control agent against *A. aurantii* and *C. dictyospermi* (Sections 1.4.2, 1.11.2). In Europe, *A. lingnanensis* is recorded from Albania, Cyprus, Greece, Italy and Spain (Noyes 2002; CABI 2007; Rasplus *et al.* 2010).

In 1964, *A. lingnanensis* from California was released in Italy (Greathead 1976). After its release, it was only recovered from one of three release sites (Greathead 1976). No information on its impact as a biocontrol agent was found.

1.56.3 *Chilocorus kuwanae* Silvestri (Col., Coccinellidae)

Chilocorus kuwanae is a predatory beetle native to Asia (Van Driesche *et al.* 1998). It is recorded from 27 species, mainly Hemiptera (including species in the families Adelgidae, Coccidae, Diaspididae and Margarodidae) (CABI 2007). Besides Europe, *C. kuwanae* was also introduced into the USA against *Adelges piceae* (Adelgidae) and into South Africa against *A. aurantii*, into India and the USA against *C. perniciosus*, into Bermuda against *P. pentagona* and into the USA against *U. euonymi* (BIOCAT 2005). In addition, it was introduced into Israel against unspecified Diaspididae and Coccidae (BIOCAT 2005). According to Roy and Migeon (2010), the species is present in Albania and Italy.

Between 1924 and 1925, *C. kuwanae* from Japan was introduced into Italy (Greathead 1976), but no information on its impact as a biological control agent was found.

1.56.4 *Chilocorus stigma* Say, twice-stabbed lady beetle (Col., Coccinellidae)

Chilocorus stigma is a predatory beetle native to Canada and the USA. Known hosts include six Diaspididae, one Aphididae (*A. kondoi*) and one Eriococcidae (*Cryptococcus fagisuga* Lindinger) (CABI 2007). Besides Europe, *C. stigma* was also introduced into Hawaii against *Asterolecanium* spp. (Asterolecaniidae), into Bermuda against *Carulaspis minima* (Signoret) and *Lepidosaphes newsteadi* (Šulc) (Diaspididae), into Australia against *C. perniciosus* (Diaspididae) and into Mauritania against *Parlatoria blanchardii* (Targioni Tozzetti) (Diaspididae) (BIOCAT 2005). In addition, it was introduced into Australia against unspecified aphids (BIOCAT 2005).

Between 1924 and 1925, *C. stigma* was introduced from Japan into Italy (Greathead 1976); no information on its impact as a biological control agent of aphids was found and the species is not recorded as present in Italy today (CABI 2007).

1.56.5 *Coccidophilus citricola* Brèthes (Col., Coccinellidae)

Coccidophilus citricola is a predatory beetle native to South America (da Silva *et al.* 2004). Known hosts include the five Diaspididae, *A. aurantii*, *D. perniciosus*, *D. echinocacti*, *L. beckii* and *P. pentagona* (CABI 2007). Besides Europe, the species was introduced into the USA against *A. aurantii* and *L. beckii*, into Bermuda against *P. pentagona* and into Hawaii against unspecified Diaspididae and Coccidae (BIOCAT 2005).

In 1937, *C. citricola* from South America was introduced into Italy, but it failed to establish there (Domenichini 1959).

1.56.6 *Comperiella bifasciata* Howard (Hym., Encyrtidae)

Comperiella bifasciata is a solitary endoparasitoid native to eastern Asia (CABI 2007). Known hosts include at least 21 Diaspididae species (Noyes 2002). In Europe, *C. bifasciata* was introduced into Cyprus, France and Greece also against *A. aurantii* and *C. dictyospermi* (Diaspididae) (Sections 1.4.4, 1.11.4). It is recorded as present in Belgium, Cyprus, the Czech Republic, France, Greece, Hungary, Italy, Moldova, the Netherlands, Romania, Russia, Spain, Ukraine and former Yugoslavia (Noyes 2002; Rasplus *et al.* 2010).

Between 1924 and 1925, 41 females and some males of *C. bifasciata* originating from the Far East were released in Italy, but according to Greathead (1976) failed to establish.

1.56.7 *Encarsia ectophaga* (Silvestri) (Hym., Aphelinidae)

Syn.: *Prospaltella ectophaga* Silvestri

Encarsia ectophaga is native to South America (Anon. 2010). Known hosts include seven Diaspididae species, i.e. *A. aurantii*, *A. hederae*, *C. dictyospermi*, *C. ficus* (Ashmead), *Hemiberlesia rapax* (Comstock), *Lepidosaphes espinosai* (Porter) and *Melanaspis paulista* (Hempel) (Anon. 2010).

In 1937, some individuals of *E. ectophaga* from South America were introduced into Italy (Greathead 1976), but apparently the species failed to establish (Noyes 2002).

1.56.8 *Pentilia egena* Mulsant (Col., Coccinellidae)

Pentilia egena is a predatory beetle native to South America (Brazil) (Guerreiro *et al.* 2003). Known hosts include seven Diaspididae, i.e. *A. destructor*, *C. aonidum*, *D. echinocacti*, *P. pergandii*, *P. cinerea* (Hadden in Doane and Hadden), *S. articulatus* and *U. citri* (Guerreiro *et al.* 2003; CABI, 2007). Besides Europe, *P. egena* was also introduced into St Lucia against *A. destructor* and into Hawaii against unspecified scale insects (BIOCAT 2005).

In 1937, *P. egena* from South America was introduced into Italy, but the species failed to establish (Domenichini 1959).

1.56.9 *Hyperaspis silvestrii* Weise (Col., Coccinellidae)

Hyperaspis silvestrii is a predatory beetle native to Africa (Schaufuß and Schenkung 1910). No information on the host range of this species was found.

Between 1924 and 1925, *H. silvestrii* originating from Japan was introduced into Italy, but it failed to establish there (Greathead 1976).

1.56.10 *Pteroptrix smithi* (Compere) (Hym., Aphelinidae)

Pteroptrix smithi is a gregarious endoparasitoid native to China (Kamburov *et al.* 1971). Two hosts are reported for this species, *C. aonidum* (L.) and *C. ficus* (Diaspididae) (Noyes 2002). Besides Europe, the species was also introduced into Israel, Mexico and South Africa against *C. aonidum* (BIOCAT 2005).

Between 1924 and 1925, *P. smithi* originating from China was released in Italy, but it failed to establish (Greathead 1976).

1.56.11 *Scymnus* sp. (Col., Coccinellidae)

Species in the genus *Scymnus* are predatory beetles.

Between 1924 and 1925, an unspecified *Scymnus* sp. originating from the Far East was released in Italy, but the species failed to establish (Greathead 1976).

1.57 Unspecified Moths (Lepidoptera)

1.57.1 *Itoplectis conquisitor* (Say) (Hym., Ichneumonidae)

Itoplectis conquisitor is a pupal parasitoid native to North America (CABI 2007). The known host range of this polyphagous parasitoid comprises at least 157 species from 104 genera (Yu *et al.* 2005), including various Lepidoptera, but also *C. rubiginosa* (Chrysomelidae; Coleoptera) and *B. curculionis* (Ichneumonidae; Hymenoptera) (CABI 2007). The species adopted *R. buoliana* Denis and Schiffermüller as host after its introduction into North America (Greathead 1976). *Itoplectis conquisitor* was also released against *R. buoliana* in Germany (Section 1.49.2).

In 1970, *I. conquisitor* was released against Lepidoptera pests in Poland (Lipa 1976), but no information on its establishment or impact as a biocontrol agent has been found.

1.58 Unspecified Coccidae Targets

1.58.1 *Metaphycus luteolus* (Timberlake) (Hym., Encyrtidae)

Metaphycus luteolus is an endoparasitoid native to North America (California) (Kapranas *et al.* 2007). It is recorded from eight Coccidae species of the genera *Coccus*, *Parthenolecanium*, *Pulvinaria* and *Saissetia* (Noyes 2002). Besides Europe, the species was also released in Australia against *C. hesperidum* (L.), in Hawaii against *C. viridis* (Green), in Bermuda against *P. psidii*, in Mauritius (Guam) against *S. coffeae* and in the USA and the former USSR against *S. oleae* (BIOCAT 2005). In Europe, *M. luteolus* is recorded from Italy, Spain and Ukraine (Noyes 2002; Rasplus *et al.* 2010).

Metaphycus luteolus was released in Italy (no date provided). According to Noyes (2002), it failed to establish, but Rasplus *et al.* (2010) reported it as established.

1.59 Unspecified Targets

1.59.1 *Mallada desjardinsi* (Navás) (Neu., Chrysopidae)

Syn.: *Mallada boninensis* Okamoto
Mallada desjardinsi is a polyphagous predator. Its native range includes the western Pacific region and the islands off the west coast of Africa. The species is commonly multiplied in commercial insectaries for augmentative releases, e.g. against citrus pests such as *Aleurocanthus* spp.

Mallada desjardinsi was released in Sicily, Italy, in the 1990s, but establishment has not yet been confirmed (A. Letardi, personal communication).

European Insect Biocontrol Agents Released in Europe

2

2.1 *Adelges piceae* (Ratzeburg), Balsam Woolly Adelgid (Hem., Adelgidae)

Syn.: *Dreyfusia piceae* (Ratzeburg)
Adelges piceae is considered by some authors to be native to the Caucasus region and introduced into Europe in the 18th century (Clausen 1978), while others consider it native to Europe but alien in Sweden and the UK (Roy and Migeon 2010). The species has further been introduced into North America and Chile (CABI 2007).

Adelges piceae colonizes *Abies alba* and other congeneric species. It attacks all epigean parts of the plants (i.e. trunk, exposed roots, branches and twigs). When heavily colonized, sap sucking by *A. piceae* and sooty mould fungi growing on honeydew often leads to the decline or death of trees. In Europe, *A. piceae* is only rarely harmful, but it has become a serious pest on many of the indigenous species of *Abies* in North America.

Chemical control of *A. piceae* is difficult as females are effectively protected by the copious wax wool covering their bodies. Insecticides sprayed from the air over forest stands were ineffective in some experiments. Ground applications of insecticide to protect Christmas trees, seed orchards and highly valued trees in parks and gardens can be effective in reducing pest population levels, but are extremely costly.

Other effective attempts to control *A. piceae* include harvesting *Abies* spp. from an infested area to protect the uninfested nearby stands, shortening of the rotation age of *Abies* spp., or conversion to non-susceptible or less susceptible *Abies* species or to other tree species (CABI 2007).

Several predatory species are reported to attack *A. piceae* (CABI 2007).

2.1.1 *Laricobius erichsonii* Rosenhauer (Col., Derodontidae)

Laricobius erichsonii is a predatory beetle native to central Europe (Clausen 1978). Adult *L. erichsonii* and later instars feed primarily on adult Adelgidae, but also on eggs. Early instars feed on eggs (Hammond and Barham 1982). This species is associated mainly with *A. piceae* but feeds also on other Adelgidae, i.e. *Adelges cooleyi*, *Adelges nordmannianae*, *Pineus pineoides* and *Pinus strobi* (Hammond and Barham 1982). *Laricobius erichsonii* has been introduced as a biological control agent for *A. piceae* and other *Adelges* spp. into Canada, Pakistan and the USA (BIOCAT 2005). In Europe, *L. erichsonii* is recorded from Belgium, Denmark, Germany, the Netherlands, Sweden, the UK, the Alps and the Carpathian Mountains (Hammond and Barham 1982).

In 1972, 1800 *L. erichsonii* from Germany (Black Forest) were introduced into the UK on an acute, not further specified *Adelges* infestation on *A. alba* (Hammond and Barham 1982). No evidence has been found that the species has become established as a result of this introduction. *Laricobius erichsonii* has, however, been reported since 1971 in other parts of the UK, presumably as a result of natural spread from localities on the coast of Belgium or the Netherlands, where it has been established for some time (Hammond and Barham 1982).

2.1.2 *Scymnus impexus* Mulsant (Col., Coccinellidae)

Scymnus impexus is a predatory beetle native throughout central and southern Europe (Clausen 1978). It is associated mainly with *A. piceae* and has often been observed to cause sharp reductions in the aphid infestations in its native distribution range (Clausen 1978). *Scymnus impexus* has been introduced against *A. piceae* into Canada and the USA (BIOCAT 2005). It has also been introduced into Pakistan against an unspecified *Adelges* species, but failed to establish there (BIOCAT 2005).

Specimens collected in southern Germany (Black Forest) were introduced into Sweden in 1968 to fight *A. piceae* (and another Adelgidae, *A. nordmannianae*). At two sites, 7000 and 12,000 specimens were released, respectively (Greathead 1976). *Scymnus impexus* established and initially responded with a parallel increase to its host population increase (Greathead 1976); however, no information on its long-term impact has been found.

In 1972, 10,000 *S. impexus* originating from southern Germany (Black Forest) were introduced into the UK, against both *A. piceae* and *A. nordmannianae* (Greathead 1976). However, the species failed to establish permanently (Duff 2008).

2.2 *Diaspidiotus perniciosus* (Comstock), San José Scale (Hem., Diaspididae)

See Section 1.18.

2.2.1 *Encarsia fasciata* (Malenotti) (Hym., Aphelinidae)

Syn.: *Prospaltella fasciata* Malenotti

Encarsia fasciata is native to Italy (Clausen 1978). Known hosts include 11 genera in the family Diaspididae (Noyes 2016). It was also introduced into Peru to control *Selenaspidus articulatus* (BIOCAT 2005). In Europe, the species is reported from Austria, France, Germany, Greece, Hungary, Italy, Montenegro, Serbia, Spain, Switzerland and the former USSR (CABI 2007).

In 1951, *E. fasciata* together with *Encarsia perniciosi* (Section 1.18.2) was imported to France to develop mass-rearing techniques (Greathead 1976). In 1958, it was re-imported and released. No information on its success as a biocontrol agent was found.

In 1956, *E. fasciata* was received and mass reared in Germany (Clausen 1978). Up to 1958, 4 million individuals had been released (Clausen 1978). No information on its success as a biocontrol agent was found.

2.3 *Dociostaurus maroccanus* (Thunberg), Moroccan Locust (Ort., Acrididae)

Dociostaurus maroccanus is a locust native to northern Africa, southern and eastern Europe, and western and central Asia. It swarms and migrates, but not over long distances (CABI 2007). Today, *D. maroccanus* is recorded from several countries in Europe, i.e. Albania, Austria, Bosnia and Herzegovina, Bulgaria, Croatia, Cyprus, France, Greece, Hungary, Italy, Former Yugoslav Republic of Macedonia, Romania and Spain (CABI 2007). While some countries are part of the native range of this species, it might have been introduced in others. For instance, it is unclear whether *D. maroccanus* is indigenous to the Carpathian Basin, and its appearance in Hungary approximately 100 years ago might have been the result of introduction from the Mediterranean region (CABI 2007).

Dociostaurus maroccanus is a highly polyphagous species, recorded from approximately 70 hosts (CABI 2007). This species prefers grass, but also frequently attacks deciduous trees, especially young oaks, feeding mainly on leaves and stems. *Dociostaurus maroccanus* can damage almost all agricultural crops, and major hosts recorded include *Avena* (oat), *Glycine max* (sojabean), *Hordeum* (barley), *Panicum miliaceum* (millet), *Secale cereale* (rye), *Triticum* (wheat) and *Zea mays* (maize) (CABI 2007). Local outbreaks in Europe have been observed in Bulgaria, Hungary and Spain (CABI 2007).

Chemical insecticides can provide control of locust and grasshopper outbreaks in general; however, the widespread use of chemicals and their associated detrimental effects on the environment, combined with the hazard they represent to users and livestock, remains a major drawback to continued reliance on their use (CABI 2007).

Control measures also include destroying egg masses, digging trenches to trap nymphs, using hopper dozers (wheeled screens) that cause locusts to fall into troughs containing water and kerosene and using poison (CABI 2007).

The entomopathogenic control agent *Metarhizium anisopliae* var. *acridum* has received considerable attention since the early 2000s as a viable biopesticide alternative to chemicals for locust control (CABI 2007).

Natural enemies recorded on *D. maroccanus* include eight parasitoid species from four genera (including many *Blaesoxipha* species (Sarcophagidae, Diptera)), five pathogen and 16 predator species from nine genera (CABI 2007). Birds and reptiles also prey on on *D. maroccanus*.

2.3.1 *Cytherea obscura* Fabricius (Dip., Bombyliidae)

Cytherea obscura is native to the Palaearctic region (Evenhuis and Greathead 1999). No information on its host range was found.

Cytherea obscura originating from mainland Italy was introduced into Sardinia in 1946 and released in six locations. An estimated 14,000 individuals including both imported *C. obscura* and *Systoechus ctenopterus* (Section 2.3.3) were released. The species established (CABI 2007), but control of *D. maroccanus* was achieved mainly by *Mylabris varabilis* (Section 2.3.2).

2.3.2 *Mylabris variabilis* (Pallas) (Col., Meloidae)

Mylabris variabilis is a predatory beetle recorded from Bulgaria, Iran, Italy, Spain and Turkey (Nikbakhtzadeh 2002; CABI 2007). Species in the subfamily Mylabrini feed on the eggs of grasshoppers.

In 1946, *M. variabilis* was introduced from mainland Italy to Sardinia, where 21,000 were released in 22 localities (Greathead 1976). The species has spread over most of the island. The impact of *M. variabilis* is considered to be partly responsible for the fact that no serious outbreaks of *D. maroccanus* have been observed in Sardinia since (Greathead 1976).

Between 1957 and 1959, 2000 specimens of *M. variabilis* were released in Corsica, France, from where complete control of the pest has been reported (OPIE 1986; Malausa *et al.* 2008).

2.3.3 *Systoechus ctenopterus* (Mikan) (Dip., Bombyliidae)

Syn.: *Systoechus sulphureus* (Mikan)

Systoechus ctenopterus is native to the Palaearctic region (Evenhuis and Greathead 1999). No information on its host range was available.

In 1946, *S. ctenopterus* originating from mainland Italy was introduced to Sardinia and released in six locations. An estimated 14,000 individuals including both imported *Systoechus sulphureus* and *C. obscura* (Section 2.3.1) were released; however, the species failed to establish (CABI 2007).

2.4 *Dialeurodes citri* (Ashmead), Citrus Whitefly (Hem., Aleyrodidae)

See Section 1.17.

2.4.1 *Encarsia tricolor* Forster (Hym., Aphelinidae)

Encarsia tricolor is an endoparasitoid native to Europe (Huang *et al.* 2009). Known hosts include nine other Aleyrodidae species in different genera, but it is also reported to be hyperparasitic on four Aphelinidae, *Encarsia formosa*, *Encarsia inaron*, *Encarsia lutea* and *Encarsia pergandiella* and the Eulophidae *Euderomphale chelidonii* (Noyes 2002; Huang *et al.* 2009). In Europe, *E. tricolor* is recorded from Belgium, Bosnia and Herzegovina, Croatia, the Czech Republic, France, Germany, Greece, Hungary, Italy, Former Yugoslav Republic of Macedonia, Montenegro, Serbia, Spain, Switzerland and the UK (Noyes 2002; CABI 2007).

In 1950, some 10,000 *Encarsia* spp., including *E. tricolor* from Belgium and Switzerland, were introduced into France (Greathead 1976); no information on its success as a biocontrol agent was found.

2.5 *Ips sexdentatus* (Boerner), Six-toothed Bark Beetle (Col., Scolytidae)

Ips sexdentatus is native to Eurasia (CABI 2007). So far, it has not been introduced to areas outside its native range, but trade will maintain a high potential for *I. sexdentatus* to become established outside its native range (CABI 2007). In fact, it has been intercepted in North America (CABI 2007). In Europe, it is currently recorded from all countries except Albania, Cyprus, Iceland, Liechtenstein and Malta (CABI 2007). In Denmark, the species has been intercepted, but has not established (CABI 2007).

Ips sexdentatus attacks conifers; major hosts include *Pinus sylvestris* and *Pinus taiwanensis* (Taiwan pine) (CABI 2007). It is also recorded from other *Pinus* species, as well as from *Abies nordmanniana* (Nordmann fir) and occasionally from *Larix* species (EPPO/CABI 1997; CABI 2007). *Ips sexdentatus* is of no significance as a pest in northern and central Europe, where it breeds only in fresh logs or in weakened or dying trees. It has, however, caused the death of *P. sylvestris* and *P. radiata* suffering from drought stress in central and southern France, northern Spain and Portugal, often in association with other pests (EPPO/CABI 1997). As the adults construct the egg galleries in the phloem under the outer bark, they defecate spores of blue stain fungi (*Ophiostoma brunneo-ciliatum* Math.). These spores germinate and the fungal hyphae grow throughout the outer sapwood, thereby blocking the movement of water from the roots to the needles. Thus, the tree is weakened by dehydration, which increases the beetles' ability to colonize a living host and ultimately cause tree death (CABI 2007). During outbreaks, large economic losses occur due to the death of trees. For instance, approximately 1 million *Picea orientalis* trees were lost in Turkey due to sporadic *I. sexdentatus* infestations (CABI 2007).

The most effective control measure is to remove infested trees before the new generation of adult beetles emerge (EPPO/CABI 1997). Unhealthy and windthrown trees, as well as slash, should be removed quickly and processed. Beetle-infested material should be cut, piled and burned. Semiochemicals are used to monitor *I. sexdentatus* populations (CABI 2007).

Many insect, mite and nematode species either consume or parasitize *Ips* spp. Natural enemies recorded on *I. sexdentatus* include 12 parasitoid species from nine genera and five predator species (CABI 2007).

2.5.1 *Aulonium ruficorne* (Olivier) (Col., Zopheridae)

Aulonium ruficorne is a predatory beetle native to Europe (Greathead 1976). It is reported as a natural enemy of Scolytidae. Besides *I. purchasi*, the species is recorded from two other Scolytidae, *Orthotomicus erosus* and *Pityogenes calcaratus* (CABI 2007). *Aulonium ruficorne* has been introduced into South Africa to control *O. erosus*

(BIOCAT 2005). In Europe, *A. ruficorne* has presumably been introduced accidentally to the UK on imported pit props from France made from maritime pine and has since established in southern UK (Greathead 1976).

In 1948, an intentional introduction of *A. ruficorne* from France was made into the UK (Greathead 1976). No information on its success as a biocontrol agent was found.

2.5.2 *Hypophloeus fraxini* Kugelann (Olivier) (Col., Tenebrionidae)

Syn.: *Corticeus fraxini* Kugelann

Hypophloeus fraxini is a predatory beetle native to Europe (Greathead 1976). It is also recorded as a natural enemy of another Scolytidae, *Hylurgops palliatus* (CABI 2007). The species has presumably been introduced accidentally to the UK on imported pit props from France (see above) and is now established in southern UK (Greathead 1976).

In 1948, an intentional introduction of *H. fraxini* was made from France into the UK (Greathead 1976). No information on the success of *H. fraxini* as a biocontrol agent was found.

2.5.3 *Platysoma oblongum* (Fabricius) (Col., Histeridae)

Platysoma oblongum is a predatory beetle native to Europe (Greathead 1976). The species is reported as a natural enemy of Scolytdiae; besides *I. sexdentatus*, the species is also recorded from *O. erosus* and *Hylurgus ligniperda* (CABI 2007). *Platysoma oblongum* has been introduced against *O. erosus* into South Africa (BIOCAT 2005).

In 1948, *P. oblongum* was introduced from France into the UK (Greathead 1976). No information on its success as a biocontrol agent was found.

2.5.4 *Tomicobia seitneri* (Ruschka) (Hym., Pteromalidae)

Tomicobia seitneri is an adult parasitoid native to Eurasia (CABI 2007). It is also recorded from other species in the genus *Ips*, i.e. *I. amitinus*, *I. duplicatus*, *I. subelongatus* and *I. typographus* (CABI 2007). The importance of *T. seitneri* on *Ips* populations is difficult to estimate; it has been observed that females can lay eggs despite being parasitized by this pteromalid (Feicht 2003).

In 1948, *T. seitneri* from Germany was introduced into the UK (Greathead 1976). No information on its success as a biocontrol agent was found.

2.6 *Liriomyza trifolii* (Burgess in Comstock), American Serpentine Leafminer (Dip., Agromyzidae)

Liriomyza trifolii originates from the Americas (Scheffer and Lewis 2006) and has been introduced into several countries in Africa, Asia, Europe, Central and South America and Oceania. It is now a major pest of the Asteraceae worldwide (CABI 2007).

In Europe, *L. trifolii* is recorded from Austria, Belgium, Croatia, Cyprus, France, Greece, Italy, Malta, Montenegro, the Netherlands, Norway, Portugal, Romania, Serbia, Slovenia, Spain and Switzerland. It has been present formerly in Bulgaria, the Czech Republic, Denmark, Finland, Germany, Hungary, Ireland, Poland, Sweden and the UK, but is today recorded as eradicated from these countries (CABI 2007).

Liriomyza trifolii is a polyphagous species recorded from 28 plant families (CABI 2007). The species attacks a wide range of ornamental and vegetable crops, including *Allium* spp. (garlic, onions, leek), beans (e.g. *G. max*, *Pisum sativum*, *Phaseolus*), celery, Chinese cabbage (*Brassica rapa* subsp. *chinensis*), chrysanthemums (*Dendranthenum* spp.), clover, *Gerbera*, lettuces, lucerne, potatoes and tomatoes. In particular, it is considered a pest on chrysanthemums (CABI 2007). *Liriomyza trifolii* attacks leaves (CABI 2007); larval mining reduces the photosynthetic ability of plants and severely infested leaves may fall, exposing plant stems to wind action and causing flower buds and developing fruit to scald (CABI 2007). In young plants and seedlings, *L. trifolii* mining can cause considerable delay in plant development, occasionally leading to plant loss (CABI 2007). The species may further act as vectors for diseases (CABI 2007). *Liriomyza trifolii* is mainly a pest of crops grown under glass, but it can also cause damage to these crops grown in the open in the warmer parts of the EPPO region (CABI 2007).

Some insecticides, particularly pyrethroids, are effective, but some individual strains of *L. trifolii* have become resistant to most insecticides, making control difficult (CABI 2007). Storage of plants at 0°C for 1–2 weeks after the eggs of *L. trifolii* hatch should kill the larvae. Gamma irradiation of eggs and first larval stages at doses of 40–50 Gy provided effective control (CABI 2007).

Foliar applications of the entomophagous nematode *Steinernema carpocapsae* reduced adult development of *L. trifolii* significantly. Biological control programmes using parasitoids have been attempted worldwide (BIOCAT 2005). Natural enemies recorded on *L. trifolii* include 47 parasitoids from 18 genera, two predators and five pathogen species.

2.6.1 *Diglyphus isaea* (Walker) (Hym., Eulophidae)

Diglyphus isaea is an ectoparasitoid of leafminers native to the Palaearctic (Eurasia) (Minkenberg 1989). The known host range includes 11 Agromyzidae species, but also three Lepidoptera species, i.e. *Pieris rapae* (Pieridae), *Plutella xylostella* (Plutellidae) and *Stigmella malella* (Nepticulidae), and the hemipteran *Lipaphis erysimi* (Aphididae) (CABI 2007). Besides Europe, *D. isaea* has also been introduced against *L. trifolii* into Canada and against another Agromyzidae, *Agromyza frontella*, into Hawaii (BIOCAT 2005). In Europe, *D. isaea* is recorded from Austria, Belgium, the Czech Republic, Finland, France, Germany, Greece, Hungary, Italy, Lithuania, the Netherlands, Malta, Poland, Spain, Sweden, the UK and former Yugoslavia (Serbia and Montenegro) (Noyes 2002).

In 1994, *D. isaea* was introduced into Malta and more than 80% parasitism was recorded (Mifsud 1997). No information on the origin of the introduced parasitoids was found; since at that time *D. isaea* was already being used as a biological control agent in Europe, it was assumed that the specimens were introduced from Continental Europe.

2.6.2 *Dacnusa sibirica* Telenga (Hym., Braconidae)

Syn.: *Pachysema sibirica* (Telenga)
Pachysema sibirica is an ectoparasitoid of leafminers native to northern Eurasia (McPartland *et al.* 2000). The known host range includes at least 12 Agromyzidae species from four genera (Yu *et al.* 2005), including leafminers such as the Agromyzidae *Liriomyza bryoniae* (Kaltenbach), *Liriomyza huidobrensis*, *L. trifolii*, *Chromatomyia horticola* and *Chromatomyia syngenesiae* (Hardy) (CABI 2007). In Europe, *P. sibirica* is recorded from Austria, Belgium, Bulgaria, Denmark, France, Germany, Hungary, Ireland, Italy, Lithuania, the Netherlands, Poland, Portugal, Spain, Sweden, the UK and former Yugoslavia (Yu *et al.* 2005; CABI 2007). As a biocontrol agent it only works as a preventative measure, since it reproduces too slowly to stop even moderate infestations (McPartland *et al.* 2000).

In 1979, *P. sibirica* was introduced into Malta (Mifsud 1997). In 1994, it was again introduced, together with *D. isaea* (Section 2.6.1), but it proved to be less effective than the latter (Mifsud 1997). No information on the origin of the introduced parasitoids was found; since at that time *D. sibirica* was already being used as a biological control agent in Europe, it was assumed that the specimens were introduced from Continental Europe.

2.7 *Prays oleae* (Bernard), Olive Kernel Borer (Lep., Yponomeutidae)

See Section 1.46.

2.7.1 *Chelonus elaeaphilus* (Silvestri) (Hym., Braconidae)

Chelonus elaeaphilus is an endoparasitoid that is thought to occur in most Mediterranean countries (Greathead 1976), although it is only recorded officially from Hungary, Italy, Portugal, Spain and former Yugoslavia (Yu *et al.* 2005). Known hosts also include five other Lepidoptera species from four families (*Coleophora hemerobiella* (Coleophoridae), *Ochromolopis staintonellus* (Epermeniidae), *Palumbina guerinii* (Gelechidae), *Pempelia turturella* and *Ephestia kuehniella* (Pyralidae)) (BIOCAT 2005; Yu *et al.* 2005).

In 1968, 1117 and 300 adult *C. elaeaphilus* from France were released at two sites in Greece (Greathead 1976). Between 1970 and 1972, an additional 6000 adults, also from France, were released (Greathead 1976). According to Greathead (1976), the species established and 26% parasitism was recorded 1 year after release at one of the 1968 release sites.

2.7.2 *Trichogramma embryophagum* (Hartig) (Hym., Trichogrammatidae)

Syn.: *Trichogramma telengai* Sorokina
Trichogramma embryophagum is an egg parasitoid native to Eurasia (CABI 2007). The species has been recorded from at least 24 Lepidoptera species, including Geometridae, Tortricidae, Pyralidae and Saturniidae (CABI 2007). It has been introduced as a biocontrol agent against *Cydia pomonella* into India (BIOCAT 2005). In Europe, it is recorded today from Bulgaria, the Czech Republic, France, Germany, Italy, Montenegro Poland, Portugal, Romania, Serbia, the Slovak Republic and the former USSR (CABI 2007).

During 1982/83, *T. embryophagum* from former Yugoslavia was released in Greece, together with other *Trichogramma* species, but only low parasitism rates were achieved (Stavraki 1985).

2.8 *Rhyacionia buoliana* (Denis and Schiffermüller), European Pine Shoot Moth (Lep., Tortricidae)

See Section 1.49.

2.8.1 *Baryscapus turionum* (Hartig) (Hym., Eulophidae)

Syn.: *Tetrastichus turionum* (Hartig)
Baryscapus turionum is an endoparasitoid native to Europe (CABI 2007). It is recorded from four Lepidoptera species, and as a hyperparasite from one Diptera (*Actia nudibasis*, Tachinidae) and one Hymenoptera species (*Copidosoma geniculatum* (Encyrtidae); Noyes 2002). *Baryscapus turionum* has also been introduced into Canada and the USA as a biocontrol agent against *R. buoliana* (BIOCAT 2005).

Between 1936 and 1937, *B. turionum* originating from Austria was introduced into the UK, where it established. The species was released simultaneously with *C. geniculatum* (Section 2.8.2). *Rhyacionia buoliana* populations collapsed due to parasitism, which presumably was due to *B. turionum* since *C. geniculatum* was a parasitoid of *Exoteleia dedecella* and not of *R. buoliana* (Greathead 1976).

2.8.2 *Copidosoma geniculatum* (Dalman) (Hym., Encyrtidae)

Copidosoma geniculatum is native to Europe (CABI 2007). The species is also recorded from *Choristoneura fumiferana* (Tortricidae) and *Exoteleia* and *Gelechia* species (Gelechiidae) (Noyes 2002). *Copidosoma geniculatum* has also been introduced into Canada and the USA as a biocontrol agent against *R. buoliana* (BIOCAT 2005).

Between 1936 and 1937, 3000 and 4400 individuals of *C. geniculatum* originating from Austria were released in the UK at two sites, respectively (Greathead 1976). The species was released simultaneously with *B. turionum* (Section 2.8.1). *Copidosoma geniculatum* was not recovered on *R. buoliana* and presumably did not attack the species (Greathead 1976).

2.9 *Scolytus scolytus* (Fabricius), Large European Elm Bark Beetle (Col., Scolytidae)

Scolytus scolytus is native to parts of Europe and Asia (CABI 2007). In Europe, *S. scolytus* is present in central, eastern and southern Europe and is recorded from all countries but Cyprus, Estonia, Finland, Iceland, Latvia, Former Yugoslav Republic of Macedonia, Malta and Norway (CABI 2007).

Scolytus scolytus lives on several elm (*Ulmus*) species. It has also been reported on *Populus nigra*, but very rarely (CABI 2007). The larvae develop in stems. Only old elms have barks thick enough to allow *S. scolytus* colonization. In the 20th century, Dutch elm disease, which was vectored by elm bark beetles (including *S. scolytus*), caused the disappearance of elms in almost all European towns.

Insecticides can be applied to kill the beetles when they arrive to feed or breed in elm trees. Both contact insecticides and systemic insecticides can be used.

To prevent and minimize *Scolytus* outbreaks, population density can be reduced by the elimination of trees already infested or suitable for colonization.

Many organisms are associated with elm bark beetles but only a limited group appears to act as predators or pathogens/parasites. The most important enemies are woodpeckers and wasps. Overall, natural enemies of *S. scolytus* include ten parasitoid species from nine genera, four predator species from three genera and five pathogen species (CABI 2007).

2.9.1 *Dendrosoter protuberans* (Nees) (Hym., Braconidae)

Dendrosoter protuberans is an ectoparasitoid native to Europe (Manojlovic *et al.* 2003). Known hosts include other Scolytidae, i.e. *Scolytus multistriatus*, *Scolytus pygmaeus*, *Scolytus ensifer*, and *Pteleobius kraatzi* (Manojlovic *et al.* 2003).

In 1971, *D. protuberans* was introduced from Austria into the UK (Greathead 1976). No information on its establishment or impact as a biocontrol agent was found.

3 Weed Biocontrol

3.1 *Ambrosia artemisiifolia* L., Common Ragweed (Asteraceae)

Common ragweed is native to Mexico and southern USA (GISD 2009). It is a summer annual plant that blooms from August to October (Brandes and Nitzsche 2007). Spread is due mostly to human activities through soil and seed transport (Bassett and Cromton 1975). Common ragweed has been introduced into Australia, Asia, Europe, New Zealand and South America (GISD 2009). In Europe, it was first recorded in the mid-1800s, but only since the early 1940s has *Ambrosia artemisiifolia* begun to spread (Juhász 1998). Multiple sources of introduction are recorded in the literature: through contamination of seeds (Chauvel *et al.* 2006), introduced via wool (Brandes and Nitzsche 2007) and during World War I with horse fodder (Chauvel *et al.* 2006). More recently, the species has also been distributed via bird seeds (Chauvel *et al.* 2006). Common ragweed is now widespread in Europe and occurs in almost all European countries (GISD 2009). It is commonly found as a ruderal plant growing in waste sites associated with frequent and extensive disturbance regimes resulting from human activities. Roadsides, railways, gravel pits, construction sites, agricultural fields, waterways, urban areas and private gardens are all sites on which this species establishes easily and grows prolifically.

Common ragweed produces highly allergenic pollen and has become the prime cause of hay fever in many European countries. The pollen causes rhinoconjunctivitis, asthma and, more rarely, contact dermatitis and urticaria. As a further consequence, tourism can be affected if visitors avoid areas with high *Ambrosia* occurrence (e.g. the Dalmatian coast). Increasingly, common ragweed has also become a major weed in agriculture, presently ranking as the number one weed in several crops in Croatia and Hungary, including sunflower (references in Brandes and Nitzsche 2007). Owing to the taxonomic closeness to sunflower, herbicide use is greatly limited, which is particularly problematic in countries such as Hungary, where sunflower is a major crop plant.

The insect complex associated with common ragweed in Europe consists mainly of polyphagous species, including some known agricultural pests. About ten species of insects, mites and fungi were recorded in Eurasia by Kovalev (1971), several generalist fungal pathogens and six insect species were found in Hungary (Bohar and Vajna 1996; Kiss *et al.* 2008) and 28 species of insects were recorded in former Yugoslavia (Maceljski and Igrc 1989).

Fungal pathogens associated with *Ambrosia* species in Eurasia are known to have a wide host range and, as observed in the case of *A. artemisiifolia*, most of them have been found to have little impact on the plant in the field (Kiss *et al.* 2003). Outbreaks of disease epidemics caused by two biotrophic fungal pathogens, *Phyllachora ambrosiae* (Berk. and M.A. Curtis) Saccardo and *Plasmophora halstedii* (Farlow) Berlese and De Toni, did affect *A. artemisiifolia* in Hungary in 1999 and 2002 (Vajna *et al.* 2000; Vajna 2002); however, for unknown reasons no similar epidemics were noted in other years (Kiss 2007). Overall, European natural antagonists are unlikely to be used to control *A. artemisiifolia* and other exotic *Ambrosia* species in Europe, leaving the classical biological control approach as the most promising option.

3.1.1 *Zygogramma suturalis* (Fabricius) (Col., Chrysomelidae)

The leaf-feeding beetle *Zygogramma suturalis* originates from Canada and the USA. Both larvae and adults feed on the leaves and flowers of common ragweed from April to mid-September. The species is generally bivoltine, occasionally univoltine or trivoltine (Igrc *et al.* 1995). It was first released as a biological control agent in 1978 in the northern Caucasus (Julien and Griffiths 1998).

In 1985, 1988 and again in 1990, *Z. suturalis* was released at three sites in former Yugoslavia, now Croatia (Igrc *et al.* 1995). Prior to its release in 1985, no-choice host specificity tests were conducted with 128 plant species/varieties, and no feeding was reported on any other plant than common ragweed (Igrc 1987). Insects for the field releases originated from the USA and from Krasnodar, Russia. In total, 2000 adults and 700 larvae were received from the USA between 1984 and 1990, and 6000 adults in 1990 from Russia (Igrc *et al.* 1995). In 1985, about 1000 larvae were released in a field near Bjelovar. Unusually low temperatures after the release were presumed to have killed most of the larvae and no permanent establishment was recorded at this site. In 1988, 250 adults were released into a field cage near Zagreb and another 250 in the open field near Zadar. At both sites, the species established and was also recorded 2 years after release. In 1990, 190 additional adults originating from the USA were released into the field cage near Zagreb. In the same year, 4000 adults originating from Russia were also released at the Zagreb site, both into field cages and into the open field.

Although *Z. suturalis* has established and has been slowly dispersing from the release sites, densities of the beetles in the field has remained low, and no control of common ragweed has been recorded so far (Igrc *et al.* 1995).

3.2 *Reynoutria japonica* var. *japonica* Houttuyn, Japanese Knotweed (Polygonaceae)

Syn.: *Fallopia japonica* var. *japonica* (Houttuyn) Ronse Decraene; *Polygonum cuspidatum* Siebold and Zuccarini

Japanese knotweed is native to Japan, Taiwan and China. It is a perennial plant, but the whole aboveground biomass dies back with the first autumn frosts. In early spring, annual stems grow from an extensive rhizome system (Beerling *et al.* 1994). A single genotype has been identified for Japanese knotweed in Europe and the species spreads almost exclusively by vegetative propagation (Krebs *et al.* 2010). Japanese knotweed has a high regeneration ability and can regenerate from small rhizome or shoot fragments (Brock and Wade 1992). Dispersal occurs mainly by transportation of soil, horticultural waste or via river systems (Child and Wade 2000). Japanese knotweed was introduced into Australia, Europe, New Zealand and North America (GISD 2009). In Europe, it was first introduced as an ornamental plant from Japan to the UK in 1825 (Seiger 1991). Japanese knotweed tolerates a wide range of habitat conditions; it is, however, particularly abundant along waterways. Naturally occurring disturbances, such as flooding, aid the transportation of rhizome and stem fragments in these habitats.

Japanese knotweed is considered among the most aggressive invasive weeds in temperate Europe and has serious consequences for biodiversity. It forms dense stands, at times even monocultures, leaving literally no space for the native vegetation and associated invertebrates. In addition to the ecological impact, Japanese knotweed can cause substantial economic damage. The stout rhizomes grow through asphalt, building foundations, concrete retaining walls and even drains, causing significant damage to infrastructure. It can add up to 10% to the costs of development and regeneration schemes.

Efforts to control knotweeds are being undertaken throughout Europe, using both mechanical and chemical methods. Control is, however, labour-intensive because the plant benefits from an extensive rhizome system and it takes several years to achieve eradication. Moreover, chemical control is limited because in most European countries the use of pesticides is restricted by law in wetlands and along rivers, i.e. in those areas invaded most heavily by alien knotweeds.

The majority of phytophagous species feeding on Japanese knotweed in Europe are generalist. The chrysomelid *Gastrophysa viridula* (De Geer), a species feeding only on members of the Polygonaceae, in particular on *Rumex* species, has been found feeding on the leaves of Japanese knotweed (Beerling *et al.* 1994); however, it cannot complete its larval development on this species (Krebs 2009). Surveys within the native range of Japanese knotweed in Japan revealed that 186 species of phytophagous arthropods were associated with the plant in its native range (Shaw *et al.* 2009).

3.2.1 *Aphalara itadori* (Shinji) (Hem., Psyllidae)

Aphalara itadori is native to Japan, Korea, Russia and the Sakhalin and Kurile Islands. Both nymphs and adults are sap suckers on the leaves and stems of Japanese knotweed from April onwards. *Aphalara itadori* is a multivoltine species. Adults overwinter presumably using tree species as shelter plants.

Prior to its release, host specificity tests were conducted using 87 plant taxa (i.e. species, subspecies and hybrids). The only plant taxa on which *A. itadori* could complete its development were exotic *Reynoutria* taxa (*Reynoutria japonica* var. *compacta*, *Reynoutria sachalinensis* and two hybrids, *Reynoutria* × *bohemica* and *Reynoutria* × *conollyana*).

Between 2010 and 2013, *A. itadori* was released in England and Wales at eight sites, with a maximum of 20,000 adults per site and year. Despite the observation of some eggs developing through to adults and small numbers of overwintered adults at some of the sites in some years, there have been no signs of establishment of *A. itadori* in the UK (Richard Shaw, Egham, 2015, personal communication).

3.3 *Cirsium arvense* (L.) Scopoli, Creeping Thistle (Asteraceae)

Creeping thistle is a perennial plant that is probably native to south-eastern Europe and the eastern Mediterranean area but has been resident throughout the rest of Europe and parts of Asia for a long time (CABI 2007). Creeping thistle has also been introduced into Africa, Australia, New Zealand and North and South America. Propagation occurs both by seeds and adventitious shoots arising from perennial roots. Creeping thistle can infest many temperate agricultural crops and is found in both disturbed (tilled) and no-tillage agricultural fields, but also in undisturbed roadsides, riverbanks, forest edges and open meadows.

Creeping thistle is a major pest and is considered one of the world's worst weeds. In Europe, it is ranked as the third most important weed and is considered an important weed in the UK, particularly in uplands and other grazing areas (Baker *et al.* 1972; CABI 2007). Creeping thistle tends to form patches when it occurs in crops and can cause substantial crop losses.

Methods for controlling creeping thistle using both non-chemical methods and herbicides have been developed for various crops. Combining herbicides with cultivation, mowing or grazing, and competitive crops has been shown to be more effective for controlling creeping thistle than herbicides alone.

At least 78 species of phytophagous insects feeding on creeping thistle are reported for Europe, including several species with high degrees of specialization on the weed (Zwölfer 1965). In the UK, however, creeping thistle does not carry the full range of phytophagous insects that attack the plant on mainland Europe. Five specialist phytophagous insects have been tested and subsequently introduced as biological control agents into Canada, New Zealand and the USA (Julien and Griffiths 1998).

3.3.1 *Altica carduorum* (Guérin-Méneville) (Col., Chrysomelidae)

Altica carduorum is a univoltine species native to southern Europe (Baker *et al.* 1972). The species is known in the field in Europe mainly from creeping thistle, with a single record of adults on *Carduus pycnocephalus* L. (Zwölfer 1965). Host specificity tests in the laboratory revealed that first-instar larvae completed their development only on *Cirsium*, *Carduus* and *Silybum* species (Harris 1964). Both larvae and adults feed externally on plants. Heavy adult feeding can cause the collapse of plants both in the laboratory and in the field (Karny 1963).

In 1969 and 1970, at least 2300 *A. carduorum* originating from France were released at four sites in the UK, both in field cages and into the environment. Some beetles overwintered successfully in cages, but no permanent establishment occurred. Egg production was low and mortality high, suggesting that *A. carduorum* was not adapted to the climatic conditions in the UK (Baker *et al.* 1972).

4 Discussion

Worldwide, more than 5000 introductions of approximately 2000 species of non-native invertebrate biological control agents (IBCAs, 'macrobials') have been made over the past 120 years in classical BC programmes, targeting arthropod pests in 196 countries or islands, and more than 1000 releases of approximately 350 species of exotic IBCAs and pathogens in classical BC programmes against weeds (Julien and Griffiths, 1998). Moreover, more than 150 species of natural enemies (parasitoids, predators and pathogens) are currently commercially available for augmentative BC (van Lenteren *et al.* 2006). Classical BC is estimated to be applied on 350 Mha worldwide (van Lenteren 2012). The comparative estimate for area usage of augmentative BC is 16 Mha (van Lenteren and Bueno 2003).

Europe has probably been the major source for natural enemies used in BC of invasive weeds and arthropods. Yet, although sometimes overlooked, deliberate releases of exotic IBCAs in Europe have a long history as well. Within the past 110 years, at least 176 exotic arthropod species have been introduced into the environment in Europe for the biological control of at least 59 insect pests and two weeds. The data on introductions into Europe are not yet in a form that enables us to generate statistics on the establishment and impact of IBCAs. Using an earlier version of the BIOCAT database, Greathead and Greathead (1992) calculated that 1445 of the 4769 records (30%) resulted in establishment and 517 (11%) achieved substantial control of the target pest. Cock *et al.* (2010) estimated that approximately one of 10–20 released natural enemy species had contributed to the reduction of the target pest.

Jacas *et al.* (2006) reviewed the introductions of IBCAs made in Spain. They listed 65 species (12 predators and 53 parasitoids) that were introduced for BC in the environment or in greenhouses. The number of introduced BC agents that were released into the environment was slightly higher than the number we reported here; this is due largely to the fact that Jacas *et al.* (2006) also include information gathered directly from colleagues involved in the introductions, while BIOCAT only includes published information. When combining the releases made into the environment and those made in greenhouses, Jacas *et al.* (2006) estimated that approximately 50% of IBCAs used in seasonal augmentative strategies and approximately 17% of IBCAs used in classical BC contributed to the reduction of the target pest.

In contrast to arthropod biological control, introductions of IBCAs for BC of invasive plants have hardly been considered in Europe. Classical biological control of weeds started more than a century ago, but up to now exotic IBCAs have been introduced against only two target weeds. An early project on weed biocontrol using exotic IBCAs in Europe was the case of bracken *Pteridium aquilinum* (L.) conducted in the UK. This research identified two specialist moths native to South Africa, *Conservula cinisigna* DE JOANNIS and *Panotima* sp. nr. *angularis*, for which host range testing was completed with 71 and 54 plant species, respectively (Lawton *et al.* 1988; Fowler, 1993). Because of potential conflicts of interest, such as bracken being considered a native species and the potential impact on species associated with bracken including at least one endangered butterfly, no releases were proposed. After a further long delay, weed biocontrol appears to have restarted in Europe; apart from the BC project against Japanese knotweed, *Reynoutria japonica* (see Section 3.2), a number of other BC projects against invasive plants are now under way (Shaw *et al.* 2011).

By far the largest numbers of IBCAs introduced into Europe are reported from countries in southern Europe (Cyprus, France, Greece, Italy and Spain). This pattern may arise from the fact that several of the IBCAs released in southern Europe had previously been released against the same target pests in other parts of the world with similar climates (e.g. in California, Australia, South Africa). For example, citrus is a crop group that originates from the Far East and that has been imported into numerous parts of the world, including the Mediterranean Basin, where conditions are ideal for the pests to thrive. Therefore, classical BC has been practised profusely in citrus-growing areas worldwide.

When releasing IBCAs for arthropod BC, potential conflicts of interest with weed BC should be taken into consideration. For example, early attempts to establish *Coccophagus montrouzeri* as a predator of *Planococcus citri* in citrus orchards in South Africa failed. However, after *Dactylopius opuntiae* was introduced for the control of invasive *Opuntia* plants, there was an abundant alternative source of prey nearby, which facilitated the establishment of *C. montrouzeri* in citrus orchards. On the other hand, along with a native ladybird (*Exochomus flaviventris*), it inhibited control of the *Opuntia* spp., and it became necessary to spray *Opuntia* infestations with insecticides to improve control by *D. opuntiae* (Annecke *et al.* 1969). The IBCA *Dirhinus giffardii*, which was introduced into Europe (see Section 1.9.3; Waterhouse 1993), may potentially cause such a conflict of interest. This parasitoid attacks mainly tephritid flies, a group of herbivores that has been used repeatedly for classical biological control of invasive weeds worldwide.

Although the first release of an exotic IBCA in Europe dates back to 1897 when the beetle *Rodolia cardinalis* was introduced to Portugal against the cottony cushion scale, *Icerya purchasi*, other regions have a much richer history in classical BC. Accordingly, the regulatory framework of the import and release of exotic organisms has lagged behind in Europe. Various guidelines and standards on implementation of the regulation of exotic IBCAs have been developed by international organizations (OECD 2003; IPPC 2005; EPPO 2010). EPPO standards are not legally binding, but they are usually followed by national authorities when implementing regulations of plant protection measures. Nevertheless, in contrast to microorganisms used as active substances in plant protection products that are regulated according to the EU Council Directive 91/414/EEC, the regulation of import and release of IBCAs in Europe is not yet harmonized (Bigler *et al.* 2006). In a report compiled as part of the EC Specific Support Action 'REBECA', Loomans (2007) summarized the current practice used for risk assessment for the import and release of IBCAs ('macrobials') in 20 countries in Europe. By 2006, eight countries (including Austria, Switzerland and the UK) had developed regulatory and administrative procedures to some degree, and six countries were still working on the design and implementation of a regulation system (including Germany, the Netherlands and Spain). Several countries (including Italy, Greece and Poland) had no regulation implemented by then. Such a lack of a harmonized regulation – and sometimes even a complete lack of regulation at the national level – in Europe can create situations where a species is not permitted for field release after an environmental risk assessment in one country, but is then released in a neighbouring country that has no regulation. The predatory ladybird *Harmonia axyridis*, which has been introduced for augmentative biological control, was first released experimentally in France and then released commercially in the Netherlands and Belgium, although it was never subject to a risk assessment at that time (Bale 2011). A retrospective analysis (van Lenteren *et al.* 2008) showed that because of its winter cold tolerance, polyphagy and dispersal potential, licences for release were not likely to have been granted. The lack of a harmonized regulation of the import and release of exotic biological control agents is particularly troublesome for classical biological control, since in this approach the biological control agents are expected to establish permanently in the environment, build up high population densities and spread actively; they will not stop at political boundaries.

In contrast to classical biological control, Europe is leading the way in commercially available augmentative biological control agents. Approximately 75% of the worldwide sales of augmentative IBCAs takes place in Europe (Cock *et al.* 2010). Until the end of the 20th century, more than half of the biological control agents used in augmentative BC (both indoors and outdoors) were exotic species. More recently, it has become common practice to look first for indigenous natural enemies when a new, even exotic, pest develops. In addition, seven exotic natural enemies that were used in Europe have recently been replaced by indigenous natural enemies (Cock *et al.* 2010).

Of the 176 IBCAs released in arthropod biological control projects in Europe, only 16 (9%) are considered to be specialized monophagous natural enemies (see Table 4). Similarly, of the 65 species introduced into Spain and released in the environment or in greenhouses, only four parasitoids are considered to be monophagous. The mean number of host species parasitized by parasitoids is 15.2, whereas the mean number of prey species attacked by predators is 21.2 (Jacas *et al.*

2006). Hence, polyphagy appears to be quite common among the BC agents that have been introduced against arthropod pests in Europe. This pattern largely corresponds with the data from arthropod biological control releases in other parts of the world. Host specificity of arthropod biological control agents became increasingly important since Howarth's (1983) seminal paper on the risks of arthropod biological control. This is in contrast to introductions made for weed biological control programmes, where since the early 1970s strong emphasis has been put on introducing highly specialized IBCAs only (Julien and Griffiths 1998).

Many of the IBCAs that were released in Europe in the past would not meet the standards of a modern environmental risk assessment, mainly because of their broad host range (see Table 4). At least in those countries that have developed a regulatory and administrative procedure regarding the import and release of IBCAs, these IBCAs would no longer be permitted for field release. There is an ongoing controversy among ecologists and environmentalists in terms of how often deliberate introductions and releases of exotic IBCAs have resulted in significant environmental impacts. Some argue that releases of exotic IBCAs have resulted in remarkably few problems (e.g. Lynch *et al.* 2001; de Clercq *et al.* 2011). Others argue that non-target effects by IBCAs have been neglected for a long time, and that those studies that explicitly assessed non-target effects by IBCAs have found some highly significant impacts, mainly by polyphagous and oligophagous IBCAs (Howarth 1991; Louda *et al.* 2003). *Harmonia axyridis* is one of the best documented examples of an IBCA that has been introduced into Europe and that impacts negatively on the environmental. Recently, Roy *et al.* (2012) have documented that this alien predator causes rapid declines of native European ladybirds. As indicated above, thorough environmental risk assessments prior to the release of a biological control agent have a shorter history in arthropod than in weed biological control projects. In the meantime, however, various recommendations have been published regarding the assessment of the environmental impact of candidates for arthropod biological control (e.g. Bigler *et al.* 2006).

Despite these concerns, it should be stressed that releasing exotic biological control agents into the environment has potentially huge ecological and economic benefits. Successful BC programmes suppressing economically important target organisms have resulted in reduced pesticide usage, reducing acute and chronic impacts of chemical pesticides on humans and the environment. The use of classical biological control has led to the protection of biodiversity and/or the protection of products harvested from natural systems, and has aided in restoring natural systems and preserving ecosystem services (van Driesche *et al.* 2010). In a recent economic assessment of 29 Australian weed biological control programmes, the annual benefit was estimated at AUS$95.3 million and the annual investment at AUS$4.3 million – a benefit to cost ratio of 23:1 (Page and Lacey 2006). The financial benefits of the use of exotic biological control agents of insect pests are less well documented, with a few notable exceptions. For example, the Alfalfa weevil, *Hypera postica* (Gyllenhal) (Coleoptera: Curculionidae), was controlled in the USA at a cost of US$1 million, producing savings of US$77 million, discounted over 32 years, and Rhodes grass scale, *Antonina graminis* (Maskell)

(Hemiptera: Pseudococcidae), was controlled in Texas for just US$200,000, giving savings of US$194 million discounted over 5 years (Greathead 1995).

In countries with a rich history of biological invasions, such as Australia and New Zealand, the increased recognition and scientific documentation of non-target effects have led to greater regulation of the import and release of exotic invertebrates for BC. It is likely that individual European countries, or the European Community as a whole, will take similar action. The challenge in the future for the regulation of the import and release of exotic IBCAs lies in improving the assessment of the associated risks, without bringing the releases of exotic IBCAs as a tool to manage arthropod pests and invasive weeds to an end. The key for sound regulation of the import and release of exotic IBCAs in the future is accurate information provided by the petitioner on the candidate IBCAs, and a careful weighing by the reviewers and regulators of the available and proposed pest control choices and their implications. The fact that Europe is a late starter in this means that it can learn from both the negative and positive experience of other countries in developing their own regulations.

Table 4. Exotic insect biocontrol agents introduced into Europe. The list includes species directly released into the environment.

Biocontrol agent	Target species in Europe	Origin[a]	Type[b]	Feeding mode[c]	Specificity[d]	Countries[e] Released	Countries[e] Recorded	Year of first introduction	Commodity
Coleoptera									
Carabidae									
Lebia grandis HENTZ	*Leptinotarsa decemlineata* (Chrysomelidae, Col.)	USA and Canada	pred		p	FR	FR	1934	Potatoes
Coccinellidae									
Ceratomegilla maculata (DE GEER)	*Hyphantria cunea* (Noctuidae, Lep.)	USA	pred		of	YU	ME, RS	1968	Broadleaved forest
Chilocorus circumdatus (GYLLENHAL IN SCHÖNHERR)	*Lepidosaphes beckii* (Diaspididae, Hem.)	India	pred		of	CY	CY	1968	Citrus
Chilocorus hauseri WEISE	*L. beckii* (Diaspididae, Hem.)	India	pred		of	CY	CY	1968/69	Citrus
Chilocorus kuwanae SILVESTRI	Unspecified scale insects (Coccidae, Hem.)	Japan	pred		p	IT	AL, IT	1924/25	?
Chilocorus similis (ROSSI)	*Diaspidiotus perniciosus* (Diaspididae, Hem.)	China	pred		of	CS	CZ, SK	1956	Fruit trees

Chilocorus stigma SAY	Unspecified scale insects (Coccidae, Hem.)	?, via Japan (NR: Canada, USA)	pred	oo	IT	–	1924/25	?
Coccidophilus citricola BRÈTHES	Unspecified scale insects (Coccidae, Hem.)	South Africa	pred	of	IT	–	1937	?
Coccinella sp.	Unspecified aphids (Aphidae, Hem.)	Hawaii	pred	–	IT	?	1908	?
Cryptolaemus montrouzieri MULSANT	*Planococcus citri* (Pseudococcidae, Hem.)	?, via USA (NR: Australia)	pred	p	CY, ES, FR, GR, IT, PT	AL, CY, ES, FR, GR, IT, NL, PL, PT, RU, SE	1908	Citrus
Harmonia axyridis PALLAS	*Aphis spiraecola*; *Toxoptera aurantii*; unspecified aphids (Aphidae, Hem.)	China, also: via FR	pred	p	FR, GR	AL, AT, BE, BG, BY, CH, CZ, DE, DK, ES, FL, FR, GR, HR, HU, IT, LI, LU, ME, MK, NL, NO, PT, RO, RS, RU, SE, SK, UA, UK	1982	Citrus
Harmonia conformis (BOISDUVAL)	*Acizzia uncatoides* (Psyllidae, Hem.)	Australia	pred	p	FR	FR	1998	Ornamental tree
Hippodamia convergens GUÉRIN-MÉNEVILLE	*Myzus persicae*; unspecified aphids (Aphididae, Hem.)	USA	pred	p	FR, IT	AL, BE, CZ, DK, SE	1908	?

Continued

Table 4. Continued.

Biocontrol agent	Target species in Europe	Origin[a]	Type[b]	Feeding mode[c]	Specificity[d]	Countries[e] Released	Countries[e] Recorded	Year of first introduction	Commodity
Nephus reunioni Fursch	*P. citri*; *Pseudococcus viburni* (Pseudococcidae, Hem.)	South Africa, also: via La Réunion	pred		oo	FR, IT	AL, ES, FR, GR, IT, PT	1974	Citrus, fruit trees
Pentilia egena Mulsant	Unspecified scale insects (Coccidae, Hem.)	South America	pred		of	IT	–	1937	?
Hyperaspis silvestrii Weise	Unspecified scale insects (Coccidae, Hem.)	?, via Japan (NR: Africa)	pred		?	IT	–	1924	?
Rhyzobius forestieri (Mulsant)	*Saissetia oleae* (Coccidae, Hem.)	?, via USA (NR: Australia)	pred		of	CY, FR, GR	AL, CY, FR, GR, IT	1980	Citrus, olive
Rhyzobius lophanthae (Blaisdell)	*D. perniciosus*; *Pseudaulacaspis pentagona* (Diaspididae, Hem.)	?, via USSR (NR: Australia)	pred		oo	CS, ES, IT	AL, CY, DE, ES, FR, GR, IT, MT, PT, UK	1908	Fruit trees, citrus
Rodolia cardinalis (Mulsant)	*Icerya purchasi* (Margarodidae, Hem.)	USA, via Egypt, FR, IT, PT	pred		og	CH, CY, ES, FR, GR, IT, MT, PT, YU	AL, BA, CH, CY, DE, ES, FR, GR, HR, IT, ME, MK, MT, PT, RS, UA, UK	1897	Fruit trees, citrus, ornamentals

Scymnus sp.	Unspecified scale insects (Coccidae, Hem.)	Far East	pred		–	IT	–	1924	?
Serangium parcesetosum SICARD	*Dialeurodes citri* (Aleyrodidae, Hem.)	? (NR: India)	pred		of	FR	FR	1985	Citrus
Monotomidae									
Rhizophagus grandis GYLLENHAL	*Dendroctonus micans* (Curculionidae, Col.)	? (NR: Siberia)	pred		oo	UK	BE, EE, FR, IT, LT, LV, UK	1983	Conifers
Staphylinidae									
Belonuchus rufipennis (FABRICIUS)	*Bactrocera oleae*; *Ceratitis capitata* (*Tephritidae*, *Diptera*)	? (NR: North and South America)	pred		p	IT	IT	1939	Olive, fruits
Diptera									
Tachinidae									
Ceromasia auricaudata TOWNSEND	*Choristoneura murinana* (Tortricidae, Lep.)	? (NR: North America)	para	endo	oo	PL	PL	1977	Fir
Panzeria ampelus WALK	*H. cunea* (Noctuidae, Lep.)	Canada, USA	para	?	of	CS, YU	–	1954	Broadleaved forest
Myiopharus doryphorae (RILEY)	*L. decemlineata* (Chrysomelidae, Col.)	? (NR: North America)	para		p	PL, IT	PL	1967	Potatoes

Continued

Table 4. Continued.

Biocontrol agent	Target species in Europe	Origin[a]	Type[b]	Feeding mode[c]	Specificity[d]	Countries[e] Released	Countries[e] Recorded	Year of first introduction	Commodity
Myiopharus sp.	*L. decemlineata* (Chrysomelidae, Col.)	? (NR: North America)	para		–	IT	?	1964	Potatoes
Smidtia fumiferanae Tothill	*C. murinana* (Tortricidae, Lep.)	? (NR: North America)	para	endo	oo	PL	PL	1977	Fir
Hemiptera									
Pentatomidae									
Perillus bioculatus (Fabricius)	*L. decemlineata* (Chrysomelidae, Col.)	?, via DE	pred	ecto	p	BE, CS, FR, DE, HU, IT, PL, MK, YU	BA, BE, CZ, DE, FR, GR, IT, HR, ME, MK, PL, RO, RS, SK	1933	Potatoes
Perilloides circumcinctus Stål	*L. decemlineata* (Chrysomelidae, Col.)	Canada	pred	ecto	p	FR	–	1933	Potatoes
Podisus maculiventris (Say)	*H. cunea* (Noctuidae, Lep.); *L. decemlineata* (Chrysomelidae, Col.)	USA	pred	ecto	p	ES, FR, PL, YU	ES, GR, IT, ME, RS	1935	Broadleaved forest, potatoes
Podisus placidus Uhler	*H. cunea* (Noctuidae, Lep.)	USA	pred	ecto	oo	YU	ME, RS	1968	Broadleaved forest

Hymenoptera									
Aphelinidae									
Aphelinus mali (HALDEMAN)	*Eriosoma lanigerum* (Aphididae, Hem.)	USA, via FR, DE, IT, Uruguay	para	endo	p	AT, BE, CH, CY, DE, DK, ES, FR, IT, MT, NL, PL, PT, SE, UK	AL, AT, BE, BG, CH, CY, CZ, DE, DK, ES, FR, HU, IT, MD, MT, NL, PL, PT, RO, SE, SK, SL, UA, UK, USSR	1920	Apple
Aphytis chrysomphali (MERCET)	Unspecified scale insects (Coccidae, Hem.)	Far East	para	ecto	of	IT	BE, CY, ES, FR, GR, HU, IT, YU	1924	?
Aphytis coheni DEBACH	*Chrysomphalus dictyospermi*; *L. beckii* (Diaspididae, Hem.)	?, via USA (NR: ?)	para	ecto	of	CY, GR	CY, GR?	1962	Citrus
Aphytis holoxanthus DEBACH	*Aonidiella aurantii* (Diaspididae, Hem.)	?, via Israel (NR: Asia)	para	endo	of	CY	BE, CY, CZ, DE, ES, FR, NL	1960	Citrus
Aphytis lepidosaphes COMPERE	*L. beckii* (Diaspididae, Hem.)	?, via FR, USA (NR: Asia)	para	ecto	of	CY, ES, FR, GR, IT	CY, ES, FR, GR, IT	1961	Citrus
Aphytis lingnanensis COMPERE	*A. aurantii*, *C. dictyospermi* (Diaspididae, Hem.); unspecified scale insects (Coccidae, Hem.)	USA	para	ecto	of	CY, ES, GR, IT	AL, CY, ES, GR, IT	1960	Citrus

Continued

Table 4. Continued.

Biocontrol agent	Target species in Europe	Origin[a]	Type[b]	Feeding mode[c]	Specificity[d]	Countries[e] Released	Countries[e] Recorded	Year of first introduction	Commodity
Aphytis melinus DeBach	*A. aurantii*; *Aspidiotus nerii*; *C. dictyospermi* (Diaspididae, Hem.)	?, via USA (NR: India and Pakistan)	para	ecto	of	CY, ES, FR, GR, IT	AL, BE, CY, CZ, DE, DK, ES, FR, GR, IT, NL, PT, YU	1961	Citrus
Aphytis proclia (Walker)	*P. pentagona* (Diaspididae, Hem.)	? (NR: Asia)	para		oo	IT	AT, BG, CH, CY, CZ, DE, ES, FR, HU, IE, ME, NL, PL, PT, RS, SK, UK, YU	1924	Mulberry
Aphytis yanonensis DeBach and Rosen	*Unaspis yanonensis* (Diaspididae, Hem.)	China	para	ecto	og	FR	FR, GR	1984	Citrus
Cales noacki Howard	*Aleurothrixus floccosus* (Aleyrodidae, Hem.)	Mexico	para		of	ES, FR, GR, IT, MT, PT	ES?, FR, GR, IT, MT, PT?	1970	Citrus
Coccobius fulvus (Compere and Annecke)	*U. yanonensis* (Diaspididae, Hem.)	?, via Japan (NR: China)	para	endo	oo	FR	FR	1984	Citrus
Coccophagus ceroplastae (Howard)	*S. oleae* (Coccidae, Hem.)	Pakistan, via FR	para	endo	oo	FR, GR	FR?, GR?	1973	Olive

Coccophagus gurneyi COMPERE	*P. citri* (Pseudococcidae, Hem.)	?, via USA (NR: Australia)	para	endo	of	IT	IT?	?	Citrus
Coccophagus rusti COMPERE	*S. oleae* (Coccidae, Hem.)	?, via USA (NR: Africa)	para		oo	GR	–	1968	Olive
Encarsia berlesei (HOWARD)	*P. pentagona* (Diaspididae, Hem.)	?, via USA, IT (NR: Japan)	para	endo	oo	AT, CH, ES, FR, IT	AL, AT, BA, BG, CH, DE, ES, FR, GR, HR, HU, IT, ME, MK, RS, USSR	1906	Mulberry
Encarsia ectophaga (SILVESTRI)	Unspecified scale insects (Coccidae, Hem.)	South America	para		of	IT	IT?	1937	?
Encarsia brimblecombei (DOZIER)	*Lepidosaphes gloverii* (Diaspididae, Hem.)	? (NR: oriental region)	para		oo	ES, FR, IT	AL, ES?, FR, IT?	1979	Fruit trees
Encarsia formosa GAHAN	*D. citri* (Aleyrodidae, Hem.)	BE, CH (NR: ?)	para	endo	of	ES, FR	AL, AT, BE, BG, CH, CS, DE, DK, ES, FI, GR, HU, IE, IT, LT, MD, MT, NL, NO, PL, PT, RO, RS, SE, USSR	1950	Citrus
Encarsia lahorensis (HOWARD)	*D. citri* (Aleyrodidae, Hem.)	?, via USA (NR: India)	para	endo	of	ES, FR, GR, IT	ES, USSR, FR, GR, IT	1973	Citrus

Continued

Table 4. Continued.

Biocontrol agent	Target species in Europe	Origin[a]	Type[b]	Feeding mode[c]	Specificity[d]	Countries[e] Released	Countries[e] Recorded	Year of first introduction	Commodity
Encarsia lounsburyi (Berlese and Paoli)	*C. dictyospermi* (Diaspididae, Hem.)	Madeira (NR: ?)	para		of	FR, IT	AL, CH, CY, ES, FR, GR, IT, NL, PT	1921	Citrus
Encarsia pergandiella Howard	*Trialeurodes vaporariorum* (Aleyrodidae, Hem.)	? (NR: New world)	para	endo	p	ES, IT	ES, FR, IT	1979	Vegetables
Encarsia perniciosi (Tower)	*C. dictyospermi*; *D. perniciosus* (Diaspididae, Hem.)	USA	para		of	AT, CH, CZ, ES, FR, DE, GR, IT, YU	AL, AT, BG, CH, CS, DE, DK, ES, FR, GR, HU, IT, PT, RO, USSR, YU	1932	Citrus, fruits
Eretmocerus debachi Rose and Rosen	*Parabemisia myricae* (Aleyrodidae, Hem.)	? (NR: Asia)	para		m	IT	IT	1991	Citrus
Eretmocerus paulistus Hempel	*A. floccosus* (Aleyrodidae, Hem.)	Mexico	para	?	m	ES	AL, ES?	1970	Citrus
Eretmocerus sp.	*P. myricae* (Aleyrodidae, Hem.)	?			–	IT	IT	1991	Citrus

Pteroptrix orientalis (SILVESTRI)	*P. pentagona* (Diaspididae, Hem.)	Japan	para	endo	of	IT	IT	1909?	?
Pteroptrix smithi (COMPERE)	Unspecified scale insects (Coccidae, Hem.)	China	para	endo	og	IT	–	1924	?
Braconidae									
Agathis unicolorata SHENEFELT	*Phthorimaea operculella* (Gelechiidae, Lep.)	?, via India (NR: South America)	para	ecto	m	CY	CY	1966	Potatoes
Aleiodes sanctihya-cinthi (GAHAN)	*H. cunea* (Noctuidae, Lep.)	Canada	para		m	CS, YU	CZ, ME, RS, SK	1954	Broadleaved forest
Apanteles fumiferanae VIERECK	*C. murinana* (Tortricidae, Lep.)	? (NR: North America)	para	endo	oo	PL	PL	1977	Fir
Apanteles scutellaris MUESEBECK	*P. operculella* (Gelechiidae, Lep.)	?, via India (NR: South America)	para	endo	of	CY	CY	1966	Potatoes
Apanteles sp.	*P. operculella* (Gelechiidae, Lep.)		para		–	GR	–	1968	Potatoes
Apanteles subandinus BLANCHARD	*P. operculella* (Gelechiidae, Lep.)	South America	para	endo	oo	CY	CY	1966	Potatoes

Continued

Table 4. Continued.

Biocontrol agent	Target species in Europe	Origin[a]	Type[b]	Feeding mode[c]	Specificity[d]	Countries[e] Released	Countries[e] Recorded	Year of first introduction	Commodity
Aphidius colemani Viereck	*A. spiraephaga*; *Diuraphis noxia*; *Melanaphis donacis* (Aphididae, Hem.)	?, via Chile, FR (NR: central Asia to Mediterranean)	para	endo	of	CZ, ES	AL, AT, BE, CH, CZ, DE, DK, ES, FI, FR, GR, HU, IE, IT, LT, MT, NL, NO, PL, PT, SE, SK, UK	1992	Wheat
Aphidius megourae Starý	*Megoura viciae* (Aphididae, Hem.)	? (NR: Russia)	para		of	CS	CS, FR, HU, PL	1962	Legumes
Aphidius smithi Sharma and Subba	*Acyrthosiphon pisum* (Aphididae, Hem.)	? (NR: India and Pakistan)	para	endo	of	CS, PL	AD, AL, AT, BE, BG, CH, CY, CZ, DE, DK, ES, FI, FR, GR, HR, HU, IE, IT, LT, MD, NL, NO, PL, PT, RU, SK, UA	1960	Legumes, lucerne
Biosteres longicaudatus Ashmead	*B. oleae* (Tephritidae, Dip.)	?, via Hawaii (NR: Taiwan, Thailand)	para	endo	p	ES, GR	GR	1952	Olive
Bracon celer Szépligeti	*B. oleae* (Tephritidae, Dip.)	Eritrea, GR: via South Africa	para	ecto	of	GR, IT	IT	1914	Olive

Habrobracon gelechiae ASHMEAD	*P. operculella* (Gelechiidae, Lep.)	?, via USA (NR: India, Pakistan)	para		p	CY, FR, MT	FR	1918	Potatoes
Chelonus spp.	*Pectinophora gossypiella* (Gelechiidae, Lep.)	Ethiopia, Australia	para		–	GR	GR?	1975	Cotton
Cotesia hyphantriae (RILEY)	*H. cunea* (Noctuidae, Lep.)	Canada	para	endo	oo	CS, YU	BG, CH, CS, DE, HU, NL, PL, YU	1954	Broadleaved forest
Fopius arisanus (SONAN)	*B. olea* (Tephritidae, Dip.)	?, via Hawaii (NR: Malesian area)	para		of	ES, GR, IT	–	1959	Fruit trees
Hymenochaonia delicata (CRESSON)	*Cydia molesta* (Tortricidae, Lep.)	USA	para	endo	oo	IT	IT	1934	Fruit trees
Lysiphlebus testaceipes (CRESSON)	*A. spiraecola*; *A. spiraephaga*; *T. aurantii*; unspecified aphids (Aphididae, Hem.)	?, via Cuba (NR: North America)	para	endo	of	CZ, ES, FR, IT, PT	AL, BA, CZ, DK, ES, FR, GR, HR, IT, MK, ME, PT, RS	1973	Citrus
Macrocentrus ancylivorus ROHWLER	*C. molesta* (Tortricidae, Lep.)	USA	para		oo	FR, IT	CH, FR?, IT	1931	Fruit trees
Meteorus bakeri COOK AND DAVIS	*H. cunea* (Noctuidae, Lep.)	Canada	para	endo	of	CS, YU	CZ, SK	1954?	Broadleaved forest
Meteorus hyphantriae RILEY	*H. cunea* (Noctuidae, Lep.)	Canada	para	endo	oo	CS, YU	CZ, SK	1954?	Broadleaved forest

Continued

Table 4. Continued.

Biocontrol agent	Target species in Europe	Origin[a]	Type[b]	Feeding mode[c]	Specificity[d]	Countries[e] Released	Countries[e] Recorded	Year of first introduction	Commodity
Chelonus blackburni Cameron	*P. gossypiella* (Gelechiidae, Lep.)	? (NR: USA)	para		oo	GR	GR	1975	Cotton
Chelonus curvimaculatus Cameron	*P. operculella* (Gelechiidae, Lep.)	South Africa	para	endo	oo	CY	CY	1965	Potatoes
Orgilus lepidus Muesebeck	*P. operculella* (Gelechiidae, Lep.)	South America	para	endo	m	CY	CY	1965	Potatoes
Orgilus parcus Turner	*P. operculella* (Gelechiidae, Lep.)	?, via India (NR: South Africa)	para	endo	of	CY	CY	1967	Potatoes
Diachasmimorpha tryoni (Cameron)	*B. oleae* (Tephritidae, Dip.)	?, via Hawaii (NR: Australia)	para		of	ES, GR	ES, GR?	1966	Olive
Pauesia cedrobii Starý and Leclant	*Cinara laportei* (Flatidae, Hem.)	Algeria, Morocco	para		m	FR	FR	1981	Cedar
Phaedrotoma trimaculata (Spinola)	*B. oleae* (Tephritidae, Dip.)	Brazil	para	?	oo	IT	IT?	1937	Olive
Phanerotoma flavitestacea Fischer	*Ectomyelois ceratoniae* (Pyralidae, Lep.)	Algeria	para	endo	oo	FR	HR, FR	1967	Carob
Psyttalia concolor Szépligeti	*B. oleae* (Tephritidae, Dip.)	Libya, Tunisia; also via IT	para	endo	of	ES, GR, FR, IT, YU	ES, FR, GR, IT	1914	Olive

Psyttalia dacicida (Silvestri)	*B. oleae* (Tephritidae, Dip.)	Eritrea	para	endo	m	IT	IT	1914	Olive
Triaspis daci (Szépligeti)	*B. oleae* (Tephritidae, Dip.)	Eritrea	para	endo	p	IT	ES	1914	Olive
Utetes africanus (Szépligeti)	*B. oleae* (Tephritidae, Dip.)	South Africa	para	endo	oo	IT	IT	1910	Olive
Chalcididae									
Dirhinus giffardii Silvestri	*C. capitata* (Tephritidae, Dip)	? (NR: eastern and western Africa)	para	endo?	oo	GR, IT	IT	1912	Fruit trees
Diapriidae									
Coptera silvestrii (Kieffer)	*C. capitata* (Tephritidae, Dip)	? (NR: western Africa)	para		og	IT	IT?	1912	Fruit trees
Dryinidae									
Neodryinus typhlocybae (Ashmead)	*Metcalfa pruinosa* (Flatidae, Hem.)	USA, via IT	para	ecto	m	CH, FR, HR, IT, SL	CH, FR, HR, IT, SL	1996	?
Encyrtidae									
Ageniaspis citricola Logvinovskaya	*Phyllocnistis citrella* (Gracillariidae, Lep.)	?, via CY, USA (NR: Thailand)	para	endo	m	CY, ES, FR, GR, IT	CY, ES?, FR?, GR?, IT, PL	1990s	Citrus
Oobius longoi (Siscaro)	*Phoracantha semipunctata* (Cerambycidae, Col.)	? (NR: Australia)	para	endo	of	PT	ES, IT, PT	1992	Eucalyptus
Coccidoxenoides peregrinus (Timberlake)	*P. citri* (Pseudococcidae, Hem.)	China	para	endo	of	IT	CY, IT	1956	Citrus

Continued

Table 4. Continued.

Biocontrol agent	Target species in Europe	Origin[a]	Type[b]	Feeding mode[c]	Specificity[d]	Countries[e] Released	Countries[e] Recorded	Year of first introduction	Commodity
Comperiella bifasciata Howard	*A. aurantii*; *C. dictyospermi* (Diaspididae, Hem.); unspecified scale insects (Coccidae, Hem.)	Far East, also via USA (NR: eastern Asia)	para	endo	of	CY, ES, FR, GR, IT	BE, CY, CZ, ES, FR, GR, HU, IT, MD, NL, RU, UA	1924	Citrus
Copidosoma koehleri Blanchard	*P. operculella* (Gelechiidae, Lep.)	?, via South Africa (NR: South America)	para	endo	oo	CY, GR, IT	AL, CY?, GR?, IT?	1947	Potatoes
Diversinervis elegans Silvestri	*S. oleae* (Coccidae, Hem.)	?, via Israel (NR: Africa)	para	endo	of	FR, GR	FR, GR	1972	Olive
Encyrtus aurantii (Geoffroy)	*S. oleae* (Coccidae, Hem.)	?, via FR (NR: Pakistan, India)	para	endo	og	FR, GR	AT, CY, CZ, ES, FR, GR, IT, USSR, YU	1953	Olive
Encyrtus fuscus Girault	*Parthenolecanium corni* (Coccidae, Hem.)	USA	para		of	IT	ES, FR, IT, PT, RO, SK, YU	1901	Fruit trees
Leptomastix dactylopii Howard	*P. citri* (Pseudococcidae, Hem.)	?, via USA (NR: South America)	para	endo	oo	CY, ES, FR, IT	AT, BA, BE, CY, CZ, DE, DK, ES, FI, FR, GR, IE, IT, NL, NO, PL, PT, SE, UK, YU	1948	Citrus

Metaphycus anneckei GUERRIERI AND NOYES/ *Metaphycus hageni* DAANE AND CALTAGIRONE	*S. oleae* (Coccidae, Hem.)	? (NR: southern Africa)	para	endo	of/m	FR?, GR, IT, PT	CY, DK, ES, FR, GR, IT, PL, PT	1971	Olive
Metaphycus helvolus (COMPERE)	*S. oleae* (Coccidae, Hem.)	?, via USA, FR (NR: southern Africa)	para	endo	oo	CY, ES, FR, GR, IT	AT, BE, CH, CY, DE, DK, ES, FR, GR, IT, NL, PT, SE, UK, USSR	1971	Olive
Metaphycus lounsburyi (HOWARD)	*S. oleae* (Coccidae, Hem.)	?, via FR, Israel (NR: South Africa)	para	endo	of	CY, ES, FR, GR, IT	AL, CY, DK, ES, FR, GR, IT, NL, PL	1973	Olive
Metaphycus luteolus (TIMBERLAKE)	Unspecified target	? (NR: North America)	para	endo	of	IT	ES, IT, UA	?	?
Metaphycus stanleyi COMPERE	*S. oleae* (Coccidae, Hem.)	?, via USA (NR: South Africa)	para	endo	of	IT	ES, GR, IT	1960	Olive
Metaphycus swirskii ANNECKE AND MYNHARDT	*S. oleae* (Coccidae, Hem.)	South Africa; also via Israel	para	endo	of	FR, GR, IT	ES, FR, GR, IT, NL	1976	Olive
Microterys nietneri (MOTSCHULSKY)	*S. oleae* (Coccidae, Hem.)	?, via USA (NR: Africa)	para	endo	of	IT	BG, CY, DE, ES, FR, HU, IT, PT, YU	1960	Olive

Continued

Table 4. Continued.

Biocontrol agent	Target species in Europe	Origin[a]	Type[b]	Feeding mode[c]	Specificity[d]	Countries[e] Released	Countries[e] Recorded	Year of first introduction	Commodity
Ooencyrtus kuwanae (Howard)	*Lymantria dispar* (Noctuidae, Lep.)	?, via Morocco, USA (NR: Japan)	para	endo	oo	CS, ES, PT	AT, BA, BG, CH, CZ, DE, ES, FR, IT, MD, PL, PT, RO, RU, RS, SK, UA, YU	1922	Deciduous trees
Psyllaephagus pilosus Noyes	*Ctenarytaina eucalypti* (Psyllidae, Hem.)	?, via USA, FR (NR: Australia)	para		m	FR, UK, IE	FR, IE, IT, UK	1994	Eucalyptus
Tetracnemoidea brevicornis (Girault)	*P. citri* (Pseudococcidae, Hem.)	? (NR: Australia)	para	endo	of	IT	FR, IT	?	Citrus
Eulophidae									
Aceratoneuromyia indica (Silvestri)	*C. capitata* (Tephritidae, Dip.)	? (NR: Indo-Pacific region)	para	endo	of	IT	IT?, UK	1909	Citrus
Cirrospilus ingenuus Gahan	*P. citrella* (Gracillariidae, Lep.)	? (NR: Asia)	para	ecto	p	CY, ES	CY?, ES?, PT	1990s	Citrus
Cirrospilus variegatus (Masi)	*B. oleae* (Tephritidae, Dip.)	Eritrea	para	ecto	p	IT	CH, HU, IT, ME, RS, SE, SK,	1914	Olive
Citrostichus phyllocnistoides (Narayanan)	*P. citrella* (Gracillariidae, Lep.)	?, via CY (NR: Asia)	para	ecto	p	CY, ES, GR	BG, CH, DE, ES, FR, GR, IT, NL, PL, RO, USSR	1990s	Citrus

Closterocerus chamaeleon (Girault)	*Ophelimus maskelli* (Eulophidae, Hym.)	? (NR: Australia)	para			IT	IT	2006	Eucalyptus
Closterocerus cinctipennis Ashmead	*Parectopa robiniella* (Gracillariidae, Lep.)	? (NR: North America)	para		p	IT	IT?	1972	Robinia
Closterocerus formosus Westwood	*B. oleae* (Tephritidae, Dip.)	Eritrea	para	endo	p	IT	AT, CH, CZ, DE, DK, ES, FI, FR, GR, HR, HU, IT, ME, NL, PL, SK, RS, SE, UK, YU	1914	Olive
Horismenus puttleri Grissell	*L. decemlineata* (Chrysomelidae, Col.)	? (NR: South America)	para	endo	og	IT		1980	Potatoes
Euderus cavasolae (Silvestri)	*B. oleae* (Tephritidae, Dip.)	Eritrea	para	ecto	m	IT	IT	1914	Olive
Galeopsomyia fausta La Salle	*P. citrella* (Gracillariidae, Lep.)	? (NR: Neotropics)	para	ecto	?	ES	–	1996	Citrus
Hyssopus thymus Girault	*Rhyacionia buoliana* (Tortricidae, Lep.)	Canada, USA	para	ecto	oo	DE	DE	1965	Pine
Quadrastichus citrella Reina and La Salle	*P. citrella* (Gracillariidae, Lep.)	Thailand, Israel	para		–	CY, ES, IT	CY, ES, IT	1990s	Citrus
Semielacher petiolatus (Girault)	*P. citrella* (Gracillariidae, Lep.)	? (NR: Australia)	para	ecto	m	CY, ES, GR, IT	CY, ES, GR?, IT, PT	1990s	Citrus

Continued

Table 4. Continued.

Biocontrol agent	Target species in Europe	Origin[a]	Type[b]	Feeding mode[c]	Specificity[d]	Countries[e] Released	Countries[e] Recorded	Year of first introduction	Commodity
Elachertus giffardi (Timberlake)	*C. capitata* (Tephritidae, Dip.)	? (NR: eastern Africa)	para	endo	of	IT	IT?	?	Fruit trees
Tetrastichus giffardianus Silvestri	*C. capitata* (Tephritidae, Dip.)	? (NR: Africa)	para	endo	of	ES	ES?	1959	Fruit trees
Thripobius javae (Girault)	*Heliothrips haemorrhoidalis* (Thripidae, Thy.)	Israel	para	endo		IT	BE, DE, DK, FR, IT, NL	1995	Fruit trees, ornamental trees
Eupelmidae									
Eupelmus afer Silvestri	*B. oleae* (Tephritidae, Dip.)	Eritrea	para	ecto	m	IT	IT?	1914	Olive
Ichneumonidae									
Campoplex haywardi Blanchard	*P. operculella* (Gelechiidae, Lep.)	South America	para	ecto	of	CY	CY	1965	Potatoes
Campoplex validus Cresson	*H. cunea* (Noctuidae, Lep.)	Canada	para		oo	CS, YU	ME, RS	1955	Broadleaved forest
Casinaria sp.	*C. murinana* (Tortricidae, Lep.)	North America	para	–	–	PL	PL?	1977	Fir
Eriborus trochanteratus (Morley)	*P. operculella* (Gelechiidae, Lep.)	India	para		oo	CY	CY	1966	Potatoes

Glypta fumiferanae (Viereck)	*C. murinana* (Tortricidae, Lep.)	? (NR: North America)	para	endo	oo	PL	PL	1977	Fir
Glypta rufiscutellaris Cresson	*C. molesta* (Tortricidae, Lep.)	? (NR: North America)	para	endo	oo	IT	IT, USSR	1934	Fruit trees
Hyposoter fugitivus (Say)	*H. cunea* (Noctuidae, Lep.)	Canada	para	endo	oo	CS, YU	BA, BG, CZ, ME, MK, SK, RS, USSR	1954	Broadleaved forest
Hyposoter pilosulus (Provancher)	*H. cunea* (Noctuidae, Lep.)	Canada	para		oo	CS	CZ, SK	1960	Broadleaved forest
Itoplectis conquisitor (Say)	*R. buoliana* (Tortricidae, Lep.); unspecified target	Canada, USA	para		p	DE, PL	DE, PL	1965	Pine
Pristomerus hawaiiensis Perkins	*P. gossypiella* (Gelechiidae, Lep.)	? (NR: ?)	para	endo	oo	GR	GR	1975	Cotton
Temelucha sp.	*P. operculella* (Gelechiidae, Lep.)	South America			–	CY	CY	1965	Potatoes
Mymaridae									
Anaphes nitens (Girault)	*Gonipterus scutellatus* (Curculionidae, Col.)	? (NR: Australia)	para	endo	of	ES, FR, IT	ES, FR, IT, PT	1993	Eucalyptus
Polynema striaticorne Girault	*Stictocephala bisonia* (Membracidae, Hem.)	USA	para	endo	of	IT	IT	1965	Fruit trees

Continued

Table 4. Continued.

Biocontrol agent	Target species in Europe	Origin[a]	Type[b]	Feeding mode[c]	Specificity[d]	Countries[e] Released	Countries[e] Recorded	Year of first introduction	Commodity
Platygastridae									
Amitus spiniferus (Brèthes)	*A. floccosus* (Aleyrodidae, Hem.)	USA, Mexico	para		m	ES, FR, IT	ES, FR?, IT?	1970	Citrus
Pteromalidae									
Halticoptera daci Silvestri	*B. oleae* (Tephritidae, Dip.)	Eritrea	para	ecto	p	IT	IT	1914	Olive
Mesopolobus modestus (Silvestri)	*B. oleae* (Tephritidae, Dip.)	Eritrea	para	ecto	m	IT	IT	1914	Olive
Moranila californica (Howard)	*S. oleae* (Coccidae, Hem.)	USA	para	endo/ecto	of	FR	ES, FR, GR, IT	1953	Olive
Scutellista caerulea (Fonscolombe)	*S. oleae* (Coccidae, Hem.)	?, via Israel (NR: Africa)	para	endo	of	FR, GR	CY, ES, FR, GR, USSR	1953	Olive
Scoliidae									
Myzinum sp.	*Melolontha melolontha* (Melolonthidae, Col.)				–	PL	PL?		
Tiphiidae									
Tiphia sp.	*M. melolontha* (Melolonthidae, Col.)				–	PL	PL?		

Torymidae									
Torymus sinensis	*Dryocosmus kuriphilus* (Cynipidae, Hym.)	Japan	para	ecto	m	IT	IT	2005	Chestnut
Trichogrammatidae									
Trichogramma cacaeciae MARCHAL	*Prays oleae* (Yponomeutidae, Lep.)	USSR, via FR	para	endo	of	GR	Native to northern Europe	1973	Citrus
Trichogramma dendrolimi MATSUMURA	*P. oleae* (Yponomeutidae, Lep.)	India	para	endo	p	GR	AT, BE, BG, BY, CH, DE, ES, FR, GR, HU, IT, LT, LV, MD, NL, PL, RO, UA, USSR	1973	Citrus
Trichogramma minutum RILEY	*Cydia pomonella* (Lep., Tortricidae; *Panolis flammea* (Lep. Noctuidae); *P. oleae* (Lep., Yponomeutidae)	Barbados, Canada, USA, also: via India	para	endo	oo	DE, GR, UK	CZ, DE, ES, FR, GR, IT, UK	1973	Olive, pine
Trichogramma pretiosum RILEY	*P. oleae* (Yponomeutidae, Lep.)	USA, YU	para	endo	oo	GR	BA?, ES, GR, HR?, ME, MK?, RS	1973	Citrus
Trichogramma spp.	*P. oleae* (Yponomeutidae, Lep.); *P. operculella* (Gelechiidae, Lep.)	USA	para		–	CY, ES, GR	ES	1944	Potatoes

Continued

Table 4. Continued.

Biocontrol agent	Target species in Europe	Origin[a]	Type[b]	Feeding mode[c]	Specificity[d]	Countries[e] Released	Countries[e] Recorded	Year of first introduction	Commodity
Neuroptera									
Chrysopidae									
Mallada desjardinsi (NAVÁS)	Citrus pests	? (NR: western Pacific region)	pred	ecto	p	IT	IT?	1990s	Citrus

[a]Country or region from where population released originates as indicated in the sources; 'via' indicates other countries where the species was released that acted as source countries for the population released; NR indicates the native range of the species.

[b]para: parasitoid; pred: predator.

[c]ecto: ectophagous; endo: endophagous.

[d]m: monophagous (only one species known as host); og: all known hosts within one genus; of: all known hosts within one family; oo: all known hosts within one order; p: known hosts from different orders.

[e]Country code abbreviations according to the International Organization for Standardization list ISO 3166 (http://www.iso.org/iso/English_country_names_and_code_elements): AD: Andorra; AL: Albania; AT: Austria; BA: Bosnia and Herzegovina; BE: Belgium; BG: Bulgaria; BY: Belarus; CH: Switzerland; CS: former Czechoslovakia; CY: Cyprus; CZ: Czech Republic; DE: Germany; DK: Denmark; EE: Estonia; ES: Spain; FI: Finland; FL: Liechtenstein; FR: France; GR: Greece; HR: Croatia; HU: Hungary; IE: Ireland; IT: Italy; LT: Lithuania; LU: Luxembourg; LV: Latvia; MD: Moldova; ME: Montenegro; MK: Macedonia; MT: Malta; NL: the Netherlands; NO: Norway; PL: Poland; PT: Portugal; RO: Rumania; RS: Serbia; RU: Russia (European part); SE: Sweden; SK: Slovak Republic; SL: Slovenia; UA: Ukraine; UK: United Kingdom; USSR: former Union of Soviet Socialist Republics; YU: former Yugoslavia.

Table 5. Insect biocontrol agents native to Europe and introduced to other parts of Europe. The list includes species directly released into the environment.

Biocontrol agent	Target species in Europe	Origin[a]	Type[b]	Feeding mode[c]	Specificity[d]	Countries[e] Released	Countries[e] Recorded	Year of first introduction	Culture
Coleoptera									
Coccinellidae									
Scymnus impexus MULSANT	*Adelges piceae* (RATZEBURG) (Adelgidae, Hemiptera)	DE	pred	ecto	m	SE, UK	SE, native to central and southern Europe	1968	Silver fir
Derodontidae									
Laricobius erichsonii ROSENHAUER	*A. piceae* (RATZEBURG) (Adelgidae, Hemiptera)	DE	pred	ecto	of	UK	BE, DE, DK, NL, SE, UK, 'Alps and Carpathian Mountains'	1972	Silver fir
Histeridae									
Platysoma oblongum (FABRICIUS)	*Ips sexdentatus* (BOERNER) (Scolytidae, Coleoptera)	FR	pred	ecto	of	UK	Native to Europe	1948	Pine
Meloidae									
Mylabris variabilis (PALLAS)	*Dociostaurus maroccanus* (THUNBERG) (Acrididae, Orthoptera)	IT	pred	ecto	oo	IT (Sardinia) FR (Corsica)	BG, IT, ES	1946	Field crops

Continued

Table 5. Continued.

Biocontrol agent	Target species in Europe	Origin[a]	Type[b]	Feeding mode[c]	Specificity[d]	Countries[e] Released	Countries[e] Recorded	Year of first introduction	Culture
Tenebrionidae									
Hypophloeus fraxini Kugelann (Olivier)	*I. sexdentatus* (Boerner) (Scolytidae, Coleoptera)	FR	pred	ecto	of	UK	Native to Europe	1948	Pine
Zopheridae									
Aulonium ruficorne (Olivier)	*I. sexdentatus* (Boerner) (Scolytidae, Coleoptera)	FR	pred	ecto	of	UK	Native to Europe	1948	Pine
Diptera									
Bombyliidae									
Cytherea obscura Fabricius	*Dociostaurus maroccanus* (Thunberg) (Acrididae, Orthoptera)	IT	para?	endo?	?	IT (Sardinia)	Native to Europe	1946	Field crops
Systoechus ctenopterus (Mikan)	*D. maroccanus* (Thunberg) (Acrididae, Orthoptera)	IT	para?	endo?	?	IT (Sardinia)	Native to Europe	1946	Field crops

Hymenoptera									
Aphelinidae									
Encarsia fasciata (MALENOTTI)	*Diaspidiotus perniciosus* (Diaspididae, Hem.)	IT	para		of	FR, GR	AT, CH, DE, ES, FR, GR, HU, IT, ME, RS, USSR	1951	Fruit trees
Encarsia tricolor FORSTER	*Dialeurodes citri* (ASHMEAD) (Aleyrodidae, Hemiptera)	BE, CH	para	endo	oo	FR	BA, BE, CH, CZ, DE, ES, FR, GR, HR, HU, IT, ME, MK, RS, UK	1950	Citrus
Braconidae									
Dacnusa sibirica TELENGA	*Liriomyza trifolii* (Agromyzidae, Dip.)	? (NR: northern Eurasia)	para	ecto	of	MT	AT, BE, BG, DE, DK, ES, FR, HU, IE, IT, LT, NL, PL, PT, SE, UK, YU	1979	Melons, cucumbers
Microchelonus elaeaphilus SILVESTRI	*Prays oleae* (BERNARD) (Yponomeutidae, Lepidoptera)	FR	para	endo	oo	GR	ES, FR?, GR?, HU, IT, PT, YU	1968	Olive
Dendrosoter protuberans (NEES)	*Scolytus scolytus* (Curculionidae, Col.)	AT	para	ecto	of	UK	Native to Europe	1971	Elm

Continued

Table 5. Continued.

Biocontrol agent	Target species in Europe	Origin[a]	Type[b]	Feeding mode[c]	Specificity[d]	Countries[e] Released	Countries[e] Recorded	Year of first introduction	Culture
Encyrtidae									
Copidosoma geniculatum (DALMAN)	*Rhyacionia buoliana* (DENIS AND SCHIFFERMULLER) (Tortricidae, Lepidoptera)	AT	para		oo	UK	Native to Europe	1936	Pine
Eulophidae									
Baryscapus turionum (HARTIG)	*R. buoliana* (DENIS AND SCHIFFERMULLER) (Tortricidae, Lepidoptera)	AT	para	endo	p	UK	Native to Europe	1936	Pine
Diglyphus isaea (WALKER)	*L. trifolii* (Agromyzidae, Dip.)	? (NR: Eurasia)	para	ecto	p	MT	AT, BE, CZ, DE, ES, FI, FR, GR, HU, IT, LT, ME, MT, NL, PL, RS, SE, UK, YU	1994	?

Pteromalidae									
Tomicobia seitneri (Ruschka)	*I. sexdentatus* (Boerner) (Scolytidae, Coleoptera)	DE	pred	ecto	og	UK	Native to Europe	1948	Pine
Trichogrammatidae									
Trichogramma embryophagum (Hartig)	*P. oleae* (Bernard) (Yponomeutidae, Lepidoptera)	YU	para	endo	oo	GR	BG, CZ, DE, FR, IT, ME, PL, PT, RO, RS, SK, USSR	1982	Olive

[a]Country or region from where population released originates, as indicated in the sources. NR indicates the native range of the species.
[b]para: parasitoid; pred: predator.
[c]ecto: ectophagous; endo: endophagous.
[d]m: monophagous (only one species known as host); og: all known hosts within one genus; of: all known hosts within one family; oo: all known hosts within one order; p: known hosts from different orders.
[e]Country code abbreviations according to the International Organization for Standardization list ISO 3166 (http://www.iso.org/iso/English_country_names_and_code_elements): AT: Austria; BA: Bosnia and Herzegovina; BE: Belgium; BG: Bulgaria; CH: Switzerland; CZ: Czech Republic; DE: Germany; DK: Denmark; ES: Spain; FI: Finland; FR: France; GR: Greece; HR: Croatia; HU: Hungary; IE: Ireland; IT: Italy; LT: Lithuania; ME: Montenegro; MK: Macedonia; MT: Malta; NL: the Netherlands; PL: Poland; PT: Portugal; RO: Rumania; RS: Serbia; SE: Sweden; SK: Slovak Republic; UK: United Kingdom; USSR: former Union of Soviet Socialist Republics; YU: former Yugoslavia.

References

Aebi A, Schönenberger N and Bigler F (2011) Evaluating the use of *Torymus sinensis* against the chestnut gall wasp *Dryocosmus kuriphilus* in the Canton Ticino, Switzerland. ART Reckenholz, Switzerland

Aguilò I (1924) La compaña contra la mosca olearia realizada por la Estación Olivarera de Tortosa. *Boletín de Agricultura técnica y económica* 16: 994–1000

Ahmad R and Muzaffar N (1976) A note on the biology of *Bracon gelechiae* (Hym.: Braconidae) and augmentation of this parasite against *Pectinophora gossypiella* (Lep.: Gelechiidae). *Entomophaga* 21: 235–238

Annecke DP, Karny M and Burger WA (1969) Improved biological control of prickly pear, *Opuntia megacantha* Salm-Dyck, in South Africa through the use of an insecticide. *Phytophylactia* 1: 9–13

Anon. (1888/9) *Insect Life*. Goverment Printing Offices, Washington DC

Anon. (2010) Universal Chalcidoidea Database. The Natural History Museum, London. Available at: http://www.nhm.ac.uk/jdsml/research-curation/research/projects/chalcidoids/index.dsml (accessed 4 May 2016)

Argyriou LC (1970) Les cochinelles de citrus an grèce. *Al Awamia* 37: 57–65

Argyriou LC (1974) Data on the biological control of citrus scales in Greece. *OILB/WPRS Bulletin* 3, 89–94

Argyriou LC (1981) Establishment of the imported parasite *Prospaltella perniciosi* (Hym.: Aphelinidae) on *Quadraspidiotus perniciosus* (Hom.: Diaspididae) in Greece. *Entomophaga* 26: 125–129

Audant P, Kreiter P, Rigollot P, Thaon M, Giuge L and Clisson S (2005) Biological control of the Japanese citrus fruit scale. *Phytoma* 583: 36–39

Audemard H, Leblon C, Neumann U and Marboutie G (1989) Bilan de sept années d'essais de lutte contre la tordeuse orientale du pêcher *Cydia molesta* Busck (Lep. Tortricidae) par confusion sexuelle des mâles. *Journal of Applied Entomology* 108: 191–207

Avidov Z, Balshin M and Gerson U (1970) Studies on *Aphytis coheni*, a parasite of the California red scale, *Aonidiella aurantii* in Israel. *BioControl* 15: 191–207

Ayouantis A (1940) Scale observed on Citrus in the island of Crete. *International Revue of Agriculture* 31: 2M–4M

Baker CRB, Blackman RL and Claridge MF (1972) Studies on *Haltica carduorum* Guerin (Coleoptera: Chrysomelidae) an alien beetle released in Britain as a contribution to the biological control of creeping thistle, *Cirsium arvense* (L.) Scop. *The Journal of Applied Ecology* 9: 819–830

Baker G and Hardy J (2005) *Black Scale Parasitoids in South Australian Olive Groves*. Entomology Unit, SARDI, Waite Campus, Adelaide, Australia, 10 pp.

Bale J (2011) Harmonization of regulations for invertebrate biocontrol agents in Europe: progress, problems and solutions. *Journal of Applied Entomology* 135: 503–513

Baklanova OV, Lappa NV and Doroshenko NN (1990) Biological investigation on the potato moth and its sensitivity to microbial pesticides. *Zashchita Rastenii (Kiev)* 37: 38–42

Barbagallo S, Longo S, Patti I and Raposarda C (1992) Efficiency of biological control against citrus whiteflies in Italy. *Bollettino di Zoologia Agraria e di Bachicoltura Ser.* II 24: 121–135

Barnes M (1991) Codling moth occurrence, host race formation and damage. In: Van der Geest L and Evenhuis H (eds) *Tortricid Pests: Their Biology, Natural Enemies and Control*. Elsevier, Amsterdam, pp. 313–327

Bartlett BR (1960) Biological races of the black scale, *Saissetia oleae*, and their specific parasites. *Annals of the Entomological Society of America* 53: 383–385

Bartoloni P (1952) La *Phthorimaea operculella* Zeller (Lep. Gelechiidae) in Italia. (Nota sulla morfologia, biologia e mezzi di lotta.) *Redia* 36: 301–379

Bassett IJ and Cromton CW (1975) The biology of Canadian weeds 11. *Ambrosia artemisiifolia* L. and *A. psilostachya* DC. *Canadian Journal of Plant Science* 55: 463–476

Beerling DJ, Bailey JP and Conolly AP (1994) Biological flora of the British Isle *Fallopia japonica* (Houtt.) Ronse Decraene (*Reynoutria japonica* Houtt.; *Polygonum cuspidatum* Sieb. and Zucc.). *Journal of Ecology* 82: 959–979

Bellows TS, Fisher TW, Caltagirone LE, Dahlsten DL, Gordh G and Huffaker CB (1999) *Handbook of Biological Control: Principles and Applications of Biological Control*. Academic Press, San Diego, California

Bénassy C and Bianchi H (1974) Observations sur *Aonidiella aurantii* MASK. et son parasite indigène *Comperiella bifasciata* How. (Hymenoptera, Encyrtidae). *Bulletin of the West Palaearctic Regional Section* 1974/3: 39–50

Bénassy C and Burgerjon A (1955) Méthode d'élevage en laboratoire de *Prospaltella perniciosi* Tow. *Annales des Epiphyties* (NS) 6: 5–10

Bénassy C, Bianchi H and Milaire H (1967) Note sur l'èfficacité en France de *Prospaltella perniciosi* Tow. *Emtomophaga* 12: 241–255

Bento A (1999) Contribuição para o estabelecimento de um programa de protecção integrada contra a traça da oliveira, *Prays oleae* (Bern.) em Trás-os-Montes. Dissert. Dout. Eng. Agric. UTAD, Vila Real, Portugal, p. 277

Berlinger MJ (1977) The Mediterranean vine mealybug and its natural enemies in southern Israel. *Phytoparasitica* 5: 3–4

Bernardo U, Viggiani G and Sasso R (2005) Biological parameters of *Thripobius semiluteus* Bouček (Hym., Eulophidae), a larval endoparasitoid of *Heliothrips haemorrhoidalis* (Bouché) (Thysan., Thripidae). *Journal of Applied Entomology* 129: 250–257

Biermann G (1973) Untersuchungen zur Einbürgerung der mordamerikanischen Pimpline *Itoplectis conquisitor* (SAY) (Hymenoptera: Ichneumonidae) als Parasit von *Rhyaconia buoliana* DEN. AND SCHIFF. (Lepidoptera: Olethreutidae) in Nordwestdeutschland. Dissertation, University of Göttingen, Germany, 134 pp.

Bigler F, Babendreier D and Kuhlmann U (2006) *Environmental Impact of Invertebrates for Biological Control of Arthropods: Methods and Risk Assessment*. CAB International, Wallingford, UK

Biliotti E, Bénassy C, Bianchi H and Milaire H (1960) Premiers essays expérimentaux d'acclimatisation en France de *Prospaltella perniciosi* TOWER, parasite spécifique importé de *Quadraspidiotus perniciosus* COMST. C. r. hebd. Séanc. *Académie d'Agriculture de France* 46: 707–711

Blomefield TL and Geertsema H (1990) First record of the oriental fruit moth, *Cydia molesta* (Lepidoptera: Tortricidae: Oletheutinae) a serious pest of peaches in South Africa. *Phytophylactica* 22: 355–357

Boardman RM (1977) Biological control: pitting insects against insects. *California Agriculture* 31: 8–11

Bodo F (1934) *Aphelinus mali* im Burgenlande heimisch? *Neuheiten Pflschutz* 27: 1–7

Bogenschütz H (1972) Freilassung von zwei nearktischen Schlupfwespenarten in der Oberrheinebene. *Mitteilungen für Naturkunde und Naturschutz* 10: 575–576

Bohar G and Vajna L (1996) Occurrence of some pathogenic fungi of common ragweed (*Ambrosia artemisiifolia* var. *elatior*) in Hungary. *Novenyvedelem* 32: 527–528 (in Hungarian)

Bonnemaison L (1965) Observations écologiques sur *Aphelinus mali* Haldemann, parasite du puceron lanigère (*Eriosoma lanigerum* Hausemann). Annales de la Société Entomologique de France (NS) 1: 143–176

Borg P (1932) *Scale Insects of the Maltese Islands*. Government Printing Office, Malta, 22 pp.

Branco M, Boavida C, Durand N, Franco JC and Mendel Z (2009) Presence of the Eucalyptus gall wasp *Ophelimus maskelli* and its parasitoid *Closterocerus chamaeleon* in Portugal: first record, geographic distribution and host preference. *Phytoparasitica* 37: 51–54

Brandes D and Nitzsche J (2007) Ecology, distribution and phytosociology of *Ambrosia artemisiifolia* L. in Central Europe. *Tuexenia* 27: 167–194

Briand LJ (1936) Shipping of potato beetle parasites and predators to France, with notes on the species involved. *Report of the Entomological Society of Ontario* 66: 31–34

Brnetić D (1968) Examination of the behaviour of the *Opius concolor* Szépl. in the ecological conditions in Central Dalmatia in 1968 (Yugoslavia), 2 pp. (typescript)

Brock JH and Wade M (1992) Regeneration of Japanese knotweed (*Fallopia japonica*) from rhizomes and stems: observation from greenhouse trials. *IX Colloque International Sur la Biologie des Mauvaise Herbes* 9: 85–93

Brown MW (1984) Literature review of *Ooencyrtus kuvanae* [Hym.: Encyrtidae], an egg parasite of *Lymantria dispar* [Lep: Lymantriidae]. *BioControl* 29: 249–265

Brown PMJ, Adriaens T, Bathon H, Cuppen J, Goldarazena A, *et al.* (2008) *Harmonia axyridis* in Europe: spread and distribution of a non-native coccinellid. *BioControl* 53: 5–21

Brudea V (2009) The efficacy of some biopesticides and vegetal metabolites in the control of Spireas aphid *Aphis spiraephaga* Müller (O: Homoptera. F. Aphidae). *Lucrări Ştiinţifice – Seria Agronomie* 52: 611–616

Brussino G, Bosio G, Baudino M, Giordano R, Ramello F and Melika G (2002) Pericoloso insetto esotico per il castagno europeo. *L'Informatore Agrario* 37: 59–61

Butcher JW and Haynes DL (1960) Influence of timing and insect biology on the effectiveness of insecticides applied for control of European pine shoot moth, *Rhyacionia buoliana*. *Journal of Economic Entomology* 53: 349–354

CABI (2007) Crop Protection Compendium. CAB International, Wallingford, UK (www.cabi.org/cpc)

Campolo O, Malacrinò A, Maione V, Laudani F, Chiera E and Palmeri V (2013) Population dynamics and spread of *Unaspis yanonensis* in Calabria, Italy. *Phytoparasitica* 41: 151–157

Campos JM, Martínez-Ferrer MT and Forés V (2006) Parasitism disruption by ants of *Anagyrus pseudococci* (Girault) and *Leptomastix dactylopii* Howard (Hymenoptera: Encyrtidae), two parasitoids of the citrus mealybug *Planococcus citri* (Risso) (Homoptera: Pseudococcidae). In: Garcia-Mari, F (ed.) *Integrated Control in Citrus Fruit Crops. Proceedings of the Meeting of the International Organization for Biological and Integrated Control of Noxious Animals and Plants*. West Palearctic Regional Section Bulletin 29, pp. 33–46

Carmichael AE, Wharton RA and Clarke AR (2005) Opiine parasitoids (Hymenoptera: Braconidae) of tropical fruit flies (Diptera: Tephritidae) of the Australian and South Pacific region. *Bulletin of Entomological Research* 95: 545–569

Carrero JM (1980) Estado actual de la lucha biologica contra las cochinillas de los agrios en Valencia (Espana). *Fruits* 35: 625–631

Cavalloro R and di Martino E (1986) *Integrated Pest Control in Citrus Groves.* Balkema, Rotterdam, the Netherlands, 600 pp.

Chaboussou F (1938) Remarques sur *Lebia grandis* Hentz. *Revue de Zoologie agriculture et appliquée* 37: 165–171

Chaboussou F (1939) Contribution à l'étude biologique de *Lebia grandis* Hentz., prédateur américain du doryphore. *Annales des Epiphyties* (NS) 5: 387–433

Charles JG (2011) Using parasitoids to infer a native range for the obscure mealybug, *Pseudococcus viburni*, in South America. *Biocontrol* 56: 155–161

Chauvel B, Dessaint F, Cardinal-Legrand C and Bretagnolle F (2006) The historical spread of *Ambrosia artemisiifolia* L. in France from herbarium records. *Journal of Biogeography* 33: 665–673

Chauzat MP, Purvis G and Dunne R (2001) A foreigner in Ireland. *Biologist* 48: 133–135

Chauzat MP, Purvis G and Dunne R (2002) Release and establishment of a biological control agent, *Psyllaephagus pilosus*, for eucalyptus psyllid (*Ctenarytaina eucalypti*) in Ireland. *Annals of Applied Biology* 141: 293–304

Child LE and Wade M (2000) *The Japanese Knotweed Manual.* Packard Publishing Limited, Chichester, UK

Chundurwar RD (1977) Life table studies and intrinsic rate of increase of *Agathis unicolorata* (Shenefelt) (Hymenoptera: Braconidae). *Plant Science* 86: 39–43

CIBC (1970) Annual report. Commonwealth Institute of Biological Control, Farnham Royal, UK.

CIP (1992) Control biologico de la polilla de la papa con *Baculovirus phthorimaea. Boletin de Capacitacion CIP* No 2

Clausen CP (1978) *Introduced Parasites and Predators of Arthropod Pests and Weeds: A World Review.* United States Department of Agriculture, Handbook No 480

Cock MJW and Perera PACR (1987) Biological control of *Opisina arenosella* Walker (Lepidoptera, Oecophoridae). *Biocontrol News and Information* 8: 283–310

Cock MJW, van Lenteren JC, Brodeur J, Barratt BIP, Bigler F, *et al* (2010) Do new access and benefit sharing procedures under the convention on biological diversity threaten the future of biological control? *BioControl* 55: 199–218

Copland MJW and Ibrahim AG (1985) Biology of glasshouse scale insects and their parasitoids. In: Hussey NW and Scopes NEA (eds) *Biological Pest Control. The Glasshouse Experience.* Blandford Press, Poole, UK, pp. 87–90

Costantino G (1935) Un nemico del contonello degli agrumi: il *Cryptolaemus montrouzieri* Muls. *Bollettini della R. Stazione Sperimentale di Frutticoltura e di Agrumicoltura di Acireale* (NS) 6, 7 pp

Cox JM (1981) Identification of *Planococcus citri* (Homoptera: Pseudococcidae) and the description of a new species. *Systematic Entomology* 6: 47–53

CSIRO (2009) *Heliothrips haemorrhoidalis.* World Thysanoptera. Available at: http://anic.ento.csiro.au/thrips/identifying_thrips/Heliothrips_haemorrhoidalis.htm (accessed 28 December 2011)

CSL (2007) CSL Pest risk analysis for *Ophelimus maskelli.* http://www.fera.defra.gov.uk/plants/plantHealth/pestsDiseases/documents/ophelimusMaskelli.pdf (accessed 4 May 2016)

Cyprus (1985) Bioecology of the olive black scale, *Saissetia oleae* (Olivier). *Annual Report, Cyprus Agricultural Research Institute* 1984: 36

da Silva RA, Michelotto MD and Busoli AC (2004) Importância de *Coccidophilus citricola* Brèthes, 1905 (Coleoptera: Coccinellidae) na predaçâo de Diaspididae no Brasil. *Revista cientifica electrônica de agronomia* 6

Daane KM and Johnson MW (2010) Olive fruit fly: managing an ancient pest in modern times. *Annual Review of Entomology* 55: 151–169

DeBach P and Argyriou LC (1967) The colonisation and success in Greece of some imported *Aphytis* spp. (Hym., Aphelinidae) parasitic on *Citrus* scale insects (Hom., Diaspididae). *Entomophaga* 12: 325–342

DeBach P, Rosen D and Kennett CE (1971) Biological control of coccids by introduced natural enemies. In: Huffacker CB (ed.) *Biological Control.* Plenum Press, New York, pp. 165–194

De Clercq P, Merlevede, F, Mestdagh, I, Vanden-Durpel, K, Mohaghegh, J and Degheele, D (1998) Predation on the tomato looper *Chrysodeixis chalcites* (Esper) (Lep., Noctuidae) by *Podisus maculiventris* (Say) and *Podisus nigrispinus* (Dallas) (Het., Pentatomidae). *Journal of Applied Entomology* 122: 93–98

De Clercq P, Mason PG and Babendreier D (2011) Benefits and risks of exotic biological control agents. *BioControl* 56: 681–698

Delanoue P (1970) Utilisation d'*Opius concolor* Szépl. en vue de la lutte contre *Dacus oleae* Gmel. (mouche de l'olive). Colloque franco-soviétique sur l'utilisation des entomophages. Antibes, 13-18 May 1968. *Annales de Zoologie Ecologie Animale, no. hors série* 63–69

Delrio G and Foxi C (2010) Current status of *Saissetia oleae* biological control in Sardinia (Italy). *IOBC/WPRS Bulletin* 59: 171–176

Delucchi VL (1976) *Studies in Biological Control.* Cambridge University, Cambridge, UK

Diaz BM, Biurrun R, Moreno A, Nebreda M and Fereres A (2006) Impact of ultraviolet-absorbing plastic films on insect vectors of virus diseases infecting crisp lettuce. *HortScience* 41: 711–716

Domenichini G. (1959) Succès et insuccès de la lutte biologique en Italie. *Transactions of the I International Conference on Insect Pathology and Biological Control Prague* 1958: 575–578

Duan JJ and Messing RH (1998) Effect of *Tetrastichus giffardianus* (Hymenoptera: Eulophidae) on nontarget flowerhead-feeding Tephritids (Diptera: Tephritidae). *Environmental Entomology* 27: 1022–1028

Duan JJ, Messing RH and Dukas R (2000) Host selection of *Diachasmimorpha tryoni* (Hymenoptera: Braconidae): comparative response to fruit-infesting and gall-forming tephritid flies. *Environmental Ecology* 29: 838–845

Dudley T, Lambert AM, Kirk A and Tamagawa Y (2008) Herbivores associated with *Arundo donax* in California. In: Julien MH, Sforza R, Bon M-C, Evans H, Hatcher P, Hinz HL and Rector BG (eds) *Proceedings of the XII International Symposium on Biological Control of Weeds.* CAB International, Wallingford, UK

Duff AG (2008) *Checklist of Beetles of the British Isles.* Wells, Somerset, UK

El Bouhssini M, Street K, Amri A, Mackay M, Ogbonnaya FC, *et al.* (2011) Sources of resistance in bread wheat to Russian wheat aphid (*Diuraphis noxia*) in Syria identified using the Focused Identification of Germplasm Strategy (FIGS). *Plant Breeding* 130: 96–97

EPPO (2005) Data sheets on quarantine pests – *Dryocosmus kuriphilus. EPPO Bulletin* 35(3): 422–424

EPPO (2010) Import and release of non-indigenous biological control agents. *EPPO Bulletin* 40: 335–344

EPPO (2014) EPPO Standards on Safe Use of Biological Control – PM 6/3 – Version 2014. List of biological control agents widely used in the EPPO region. Available at: http://archives.eppo.int/EPPOStandards/biocontrol_web/bio_list.htm#classic (accessed May 2016)

EPPO/CABI (1997) Data Sheets on Quarantine Pests: *Ips sexdentatus*. Available at: https://www.eppo.int/QUARANTINE/data_sheets/insects/IPSXSE_ds.pdf (accessed 25 May 2016)

Evans EW (1982) Habitat differences in feeding habits and body size of the predatory stinkbug *Perillus circumcinctus* (Hem., Pentatomidae). *New York Entomological Society* 90: 129–133

Evenhuis HH (1962) Betrachtungen über den Einfluss der Blutlauszehrwespe *Aphelinus mali* (HALD.) auf den Massenwechsel ihres Wirtes, der Apfelblutlaus *Eriosoma lanigerum* (HAUSM.) in den Niederlanden. *Zeitschrift für angewandte Entomologie* 49: 402–407

Evenhuis NL and Greathead DJ (1999) *World Catalog of Bee Flies (Diptera: Bombyliidae)*. Backhuys Publishers, Leiden, the Netherlands, 756 pp

Fabre JP and Rabasse JM (1987) Introduction dans le Sud-Est de la france d'un parasite: *Pauesia cedrobii* [Hym.: Aphidiidae] du puceron: *Cedrobium laportei* [Hom.: Lachnidae] du cèdre de l'atlas: *Cedrus atlantica*. *BioControl* 32: 127–141

Faes H. (1928) Un contre-parasite du puceron lanigère, l'*Aphelinus mali*. *Annuaire agricole de la Suisse* 29: 515–519

Feicht E (2003) Erz- und Brackwespen. *LWF aktuell* 38: 9–11

Fenili GA and Pegazzano F (1962) Esperimento di introduzione inToscana di *Opius siculus* MONASTERO (*concolor* SZÉPL.) parassita del *Dacus oleae* GMEL. *Redia* 47: 173–187

Feron M, Bernard R and Poitout S (1961) La mouche de l'olive, *Dacus oleae* Gmel., et ses parasites en Corse en 1959 et 1960. *Entomophaga* 6: 173–183

Ferriera L (1939) A luta contra *Pseudococcus citri* RISSO e o problema general da luta biologica. *Palestras agronomicos* 2: 17–47

Fielding NJ and Evans HF (1997) Biological control of *Dendroctonus micans* (Scolytidae) in Great Britain. *Biocontrol News and Information* 18: 51N–60N

Fischer M and Koponen M (1999) A survey of Opiinae (Hymenoptera, Braconidae) of Finland, part 1. *Entomologica Fennica* 10: 65–93

Fodor E and Hâruta O (2009) Niche partition of two invasive insect species, *Parectopa robiniella* (Lepidoptera: Gracillariidae) and *Phyllonorycter robiniella* (Clem.) (Lepidoptera: Gracillariidae). *Research Journal of Agricultural Science* 41: 261–269

Fowler SV (1993) The potential for control of bracken in the UK using introduced herbivorous insects. *Pesticide Science* 37: 393–397

Fox PM, Pass BC and Thurston R (1967) Laboratory studies on the rearing of *Aphidius smithi* (Hyinenoptera: Braconidae) and its parasitism of *Acyrthosiphon pisum* (Homoptera: Aphididae). *Annals of the Entomological Society of America* 60: 1083–1087

Frank JH and Barreray R (2010) Natural history of *Belonuchus* NORDMANN spp. and allies (Coleoptera: Staphylinidae) in *Heliconia* L. (Zingiberales: Heliconiaceae) flower bracts. *Insecta Mundi* 0110: 1–12

Franz JM (1966) Populationsdynamik und biologische Bekämpfung des Kartoffelkäfers. *Entomophaga* 11: 470

Furuhashi K and Nishino M (1983) Biological control of arrowhead scale, *Unaspis yanonensis*, by parasitic wasps introduced from the People's Republic of China. *BioControl* 28: 277–286

Gambaro PI (1965) Osservazione sulla biologia della *Prospaltella perniciosi* Tow. nella valle Padana. *Entomophaga* 10: 373–376

Garcia-Mari F, Vercher R, Costa-Comelles J, Marzal C and Villalba M (2003) Establishment of *Citrostichus phyllocnistoides* (Hymenoptera: Eulophidae) as a biological control agent for the citrus leafminer *Phyllocnistis citrella* (Lepidoptera: Gracillariidae), in Spain. *Biological Control* 29: 215–226

Geier P and Baggiolini M (1950) Quelques observations sur la biologie de *Pericerya purchasi* Mask. au Tessin (Homopt., Margarodid.). *Mitteilungen der schweizerischen entomologischen Gesellschaft* 23: 104–116

Gelmroth KG (1972) Aufzucht- und Aussetzungsversuche mit dem nordamerikanischen Parasiten *Elachertus (Hyssopus) thymus* Girault (Hym. chalcidoidea, Eulophidae) zur biologischen Bekämpfung des Kiefernknospentreibwicklers *Rhyacionia buoliana* Schiff. (Lep. Tortricid.). *Anzeiger für Schädlingskunde und Pflanzenschutz* 45: 49–52

Genduso P and Ragusa S (1968) Lotta biologica artificiale contro la mosca delle olive a mezzo dell'*Opius c. siculus* Mon. in Puglia nel 1968. *Bollettino dell'Istituto di entomologia agraria e dell'Osservatorio di fitopatologia di Palermo* 7: 196–216

Gibbs M, Schönrogge K, Alma A, Melika G, Quacchia A, *et al.* (2011) *Torymus sinensis*: a viable management option for the biological control of *Dryocosmus kuriphilus* in Europe? *BioControl* 56: 527–538

Gilbert M and Gregoire JC (2003) Site condition and predation influence a bark beetle's success: a spatially realistic approach. *Agricultural and Forest Entomology* 5: 87–96

GISD (2009) Global Invasive Species Database. *Ambrosia artemisiifolia*. Available at: http://www.iucngisd.org/gisd/species.php?sc=1125 (accessed May 2016)

Gómez Clemente F (1954a) Los insectos auxillares en la luncha contra los nocivos a los agrios. Algunos estudios realizados en la Estación de Fitzopatologia Agricola de Burjasot (Valencia). *Boletín de patología vegetal y entomología agrícola* 19: 1–18

Gómez Clemente F (1954b) Estado actual de la luncha biológica contra algunas cochinillas de los agrios (*Pseudococcus citri* y *Pericerya purchasi*). *Boletín de patología vegetal y entomología agrícol* 19: 19–35

Greathead DJ (1971) A review of biological control in the Ethiopian Region. *Technical Communication Commonwealth Institute of Biological Control* 5: 162 pp

Greathead DJ (1976) *A Review of Biological Control in Western and Southern Europe*. Commonwealth Agricultural Bureaux Slough, UK

Greathead DJ (1995) Benefits and risks of classical biological control. In: Hokkanen HMT and Lynch JM (eds) *Biological Control: Benefits and Risks*. Cambridge University Press, Cambridge, UK, pp. 53–63

Greathead DJ and Greathead AH (1992) Biological control of insect pests by insect parasitoids and predators: the BIOCAT database. *Biocontrol News and Information* 13: 61N–68N

Greenberg SM, Jones WA and Liu T-X (2008) *Bemisia tabaci* (Homoptera: Aleyrodidae) instar effects on rate of parasitism by *Eretmocerus mundus* and *Encarsia pergandiella* (Hymenoptera: Aphelinidae). *Entomological Science* 11: 97–103

Guerreiro JC, Busoli AC and Filho EB (2003) Oviposition and predation of *Pentilia egena* Mulsant (Coleoptera: Coccinellidae) in response to temperature. *Scientia Agricola (Piracicaba, Brazil)* 60: 587–589

Guerrieri E and Noyes JS (2000) Revision of European species of genus *Metaphycus* Mercet (Hymenoptera: Chalcidoidea: Encyrtidae), parasitoids of scale isects (Homoptera: Coccoidea). *Systematic Entomology* 25: 147–222

Györfi J (1956) Die in den Maikäfer- und anderen Blatthornkäferlarven schmarotzenden Wespen. *Zeitschrift für Angewandte Entomologie* 38: 468–474

Hamlet J (2005) MEALYBUG: Why we should sit up and take notice. Hawke's Bay Focus Vineyard Project

Hammond PM and Barham CS (1982) *Laricobius erichsoni* Rosenbauer (Coleoptera: Derodontidae). A species and superfamily new to Britain. *Entomologist's Gazette* 33: 35–40

Hanks LM, Paine TD and Millar JG (1996) Tiny wasp helps protect eucalypts from eucalyptus longhorned borer. *California Agriculture* 50: 14–16

Harris P (1964) Host specificity of *Altica carduorum* Guer. (Coleoptera: Chrysomelidae). *Canadian Journal of Zoology* 42: 857–862

Hayat M (1989) A revision of the species of *Encarsia foerster* (Hymenoptera: Aphelinidae) from India and adjacent countries. *Oriental Insects* 23: 1–131

Heinrich C (1956) *American Moths of the Subfamily Phycitinae*. United States National Museurn Bulletin 207. Smithsonian lnstitution, Washington, DC, 581 pp.

Heraty J, Woolley J and Polaszek A (2007) Catalog of the *Encarsia* of the World. Available at: http://cache.ucr.edu/~heraty/Encarsia.cat.pdf (accessed 4 May 2016)

Hill DS (2008) *Pests of Crops in Warmer Climates and Their Control*. Springer, Berlin

Hoddle MS, Van Driesche RG and Sanderson JP (1998) Biology and use of the whitefly parasitoid *Encarsia formosa*. *Annual Review of Entomology* 43: 645–669

Hodkinson ID (1999) Biocontrol of eucalyptus psyllid *Ctenarytaina eucalypti* by the Australian parasitoid *Psyllaephagus pilosus*: a review of current programmes and their success. *Biocontrol News and Information* 20: 129–134

Hokkanen HMT and Pimentel D (1989) New associations in biological control: theory and practice. *Canadian Entomologist* 121: 828–840

Howard LO (1929) *Aphelinus mali* and its travels. *Annals of the Entomological Society of America* 22: 311–368

Howard LO, Gill TN, Schwarz EA, Ashmead WH and Chittenden FH (1901) *Proceedings of the Entomological Society of Washington*, 1896–1901. Entomological Society of Washington

Howarth FG (1983) Classical biocontrol: panacea or Pandora's box? *Proceedings of the Hawaiian Entomological Society* 24: 239–244

Howarth FG (1991) Environmental impacts of classical biological control. *Annual Review of Entomology* 36: 485–509

Hoy MA and Nguyen R (1997) Classical biological control of the citrus leafminer *Phyllocnistis citrella* Stainton (Lepidoptera: Gracillaridae): theory, practice, art and science. *Tropical Lepidoptera* 8: 1–19

Huang Y, Loomans AJM, van Lenteren JC and RuMei X (2009) Hyperparasitism behaviour of the autoparasitoid *Encarsia tricolor* on two secondary host species. *BioControl* 54: 411–424

Igrc J (1987) Proucavanje zlatice *Zygogramma suturalis* Farbricius (Coleoptera, Chrysomelidae) potencijalnog agensa bio. *Agriculturae Conspectus Scientificus* 76/77: 31–56

Igrc J, DeLoach CJ and Zlof V (1995) Release and establishment of *Zygogramma suturalis* F. (Coleoptera: Chrysomelidae) in Croatia for control of common ragweed (*Ambrosia artemisiifolia* L.). *Biological Control* 5: 203–208

Inserra S (1966) Introduzione et acclimatazione di due *Aphytis* (*A. melinus* DEBACH et *A. lingnanensis* COMPERE) parassiti ectofagi di alcune cocciniglie degli agrumi. *Tecnica Agricola* 18: 176–186

Inserra S (1968) Prove di lotta integrata contro l'*Aonidiella aurantii* MASK. *et al.* tre cocciniglie degli agrumi in Sicilia. *Entomologica* 4: 45–77

Inserra S (1970) Acclimatation, diffusion et notes sur la biologie d'*Aphytis melinus* DEBACH en Sicilie. *Al Awamia* 37: 39–55

Iperti G, Giuge L and Roger JP (1989) Installation de *Rhyzobius forestieri* (Col., Coccinellidae) sur l'Ile de Porquerolles. *Entomophaga* 34: 365–372

IPPC (2005) Guidelines for the export, shipment, import and release of biological control agents and other beneficial organisms. International Standards for Phytosanitary Measures No 3. Available at: https://www.ippc.int/en/publications/600/ (accessed 25 May 2016)

Ivanova TS, Borovaya VP and Danilov LG (1994) A biological method of controlling the potato moth. *Zashchita Rastenii (Moskva)* No 2: 39

Jacas J-A, Urbaneja A and Viñuela E (2006) History and future of introduction of exotic arthropod biological control agents in Spain: a dilemma? *BioControl* 51: 1–30

Jackson CG, Delph JS and Neemann EG (1978) Development, longevity and fecundity of *Microchelonus blackburni* (Hym.: Braconidae) as a parasite of *Pectinophora gossypiella* (Lep.: Gelechidae). *BioControl* 23: 35–42

Jamieson LE, Froud K, Edwards R and Stevens PS (2008) Establishment of *Thripobius javae* (= *semiluteus*) in New Zealand. *New Zealand Plant Protection* 61: 17–23

Jiménez A, Arroyo M and Mellado L (1969) Ensayos preliminares de introducción en España de *Opius concolor* Szépl., parásito endófago de *Dacus oleae* Rossi. *Boletín de patología vegetal y entomología agrícola* XXXL: 29–32

Juhász M (1998) *The History of Ragweed in Europe*. 6th International Congress on Aerobiology, Satellite Symposium Proceedings: Ragweed in Europe. Alk Abelló, Perugia, Italy

Julien MH and Griffiths MW (1998) *Biological Control of Weeds. A World Catalogue of Agents and their Target Weeds*, Fourth Edition. CAB International, Wallingford, UK

Kamburov S, Gerling D and Bar D (1971) Mass rearing, distribution and colonization of *Pteroptrix smithi* (Compere) in Israel. *BioControl* 15: 413–418

Kapranas A, Morse JG, Forster LD and Luck RF (2007) Survey of brown soft scale *Coccus hesperidum* L. parasitoids in southern California citrus. *Biological Control* 42: 288–299

Karamaouna F and Copland MJK (2000) Host suitability, quality and host size preference of *Leptomastix epona* and *Pseudaphycus flavidulus*, two endoparasitoids of the mealybug *Pseudococcus viburni*, and host size effect on parasitoid sex ratio and clutch size. *Entomologia Experimentalis et Applicata* 96: 149–158

Karny M (1963) *Haltica carduorum* Guer. (Col.: Chrysomelidae) attacking *Cirsium arvense*. *Technical Bulletin Commonwealth Institute of Biological Control* 3: 99–110

Katsoyannos P (1984) The establishment of *Rhizobius forestieri* (Col. Coccinellidae) in Greece and its efficiency as an auxilliary control agent against a heavy infestation of *Saissetia oleae* (Hom. Coccidae). *Entomophaga* 29: 387–398

Katsoyannos P, Ifantis K and Kontodimas DC (1997) Phenology, population trend and natural enemies of *Aleurothrixus floccosus* (Hom.: Aleyrodidae) at a newly invaded area in Athens, Greece. *Entomophaga* 42: 619–628

Keasara T and Sadeh A (2007) The parasitoid *Copidosoma koehleri* provides limited control of the potato tuber moth, *Phthorimaea operculella*, in stored potatoes. *Biological Control* 42: 55–60

Kfir R (1997) The diamondback moth with special reference to its parasitoids in South Africa. In: Sivapragasam A, Loke WH, Hussan AK and Lim GS (eds) *Management of Diamondback Moth and other Crucifer Pests*. Malaysian Agricultural Research and Development Institute

Klett WEW (1959) Die Arbeiten zur Zucht und Freilassung von *Prospaltella perniciosi* TOWER (Hymenoptera) in Europa. *Verhandlungen des 4. Internationalen Pflanzenschutz-Kongresses* 1957 1: 861–862

Kiss B, Rédei D and Koczor S (2008) Occurrence and feeding of hemipterans on common ragweed (*Ambrosia artemisiifolia*) in Hungary. *Bulletin of Insectology* 61: 195–196

Kiss L (2007) Why is biocontrol of common ragweed, the most allergenic weed in Eastern Europe, still only a hope? In: Vincent C, Goettel MS and Lazarovits G (eds) *Biological Control: A Global Perspective*. CAB International, Wallingford, UK, pp. 80–90

Kiss L, Vajna L, Bohár G, Varga K, Paksiri U, *et al.* (2003) *Phyllachora* epidemic on common ragweed (*Ambrosia artemisiifolia*): a unique natural control phenomenon in Hungary in 1999. Workshop Biocontrol of Weeds with Pathogens, Lincoln, New Zealand, pp. 17–18

Koehler CS, Kattoulas ME and Frankie GW (1966) Biology of *Psylla uncatoides*. *Journal of Economic Entomology* 59: 1097–1100

Kovalev OV (1971) Modern outlooks of biological control of weed plants in the U.S.S.R. and the international phytophagous exchange. In: Dunn PH (ed.) *Second International Symposium on Biological Control of Weeds*, Rome, Italy. Commonwealth Agricultural Bureaux, Farnham Royal, Slough, UK, pp. 166–172

Krambias A and Kontzonis A (1980) Establishment of *Leptomastix dactylopii* (How.) in Cyprus. *Fruits* 35: 783–785

Krebs C (2009) Hybridization in the invasive *Fallopia* complex and its influence on sexual reproduction and herbivore resistance. Dissertation, Philipps-Universität, Marburg, Germany

Krebs C, Mahy G, Matthies D, Schaffner U, Tiébré M-S and Bizoux J-P (2010) Taxa distribution and RAPD markers indicate different origin and regional differentiation of hybrids in the invasive *Fallopia* complex in central-western Europe. *Plant Biology* 12: 215–23

Laamari M, Jousselin E and Coeur d'Acier A (2010) Assessment of aphid diversity (Hemiptera: Aphididae) in Algeria: a fourteen-year investigation. *Entomologie faunistique – Faunistic Entomology* 62: 73–87

Lacey, LA and Unruh, TR (2007) Biological control of codling moth (*Cydia pomonella* Tortricidae: Lepidoptera) and its role in integrated pest management, with emphasis on entomopathogens. *Vedalia* 12, 33–60

Lacey LA, Thomson D, Vincent C and Arthurs SP (2008) Codling moth granulovirus: a comprehensive review. *Biocontrol Science and Technology* 18: 639–663

Lambert, AM, Dudley, TL and Saltonstall, K (2010) Ecology and impacts of the large-statured invasive grasses *Arundo donax* and *Phragmites australis* in North America. *Invasive Plant Science and Management* 3: 489–494

La Salle J and Pefia JE (1997) A new species of *Galeopsomyia* (Hymenoptera, Eulophidae: Tetrastichinae): a fortuitous parasitoid of the citrus leafminer, *Phyllocnistis citrella* (Lepidoptera, Gracillariidae). *Florida Entomologist* 80: 461–470

Lawton JH, Rashbrook VK and Compton SG (1988) Biocontrol of British bracken: the potential of two moths from Southern Africa. *Annals of Applied Biology* 112: 479–490

Le Berre, JR and Portier, G (1963) Utilisation d'un hétéroptère Pentatomidae *Perillus bioculatus* (FABR.) dans la lutte contre le doryphore *Leptinotarsa decemlineata* (SAY): premiers résultats obtenus en France. *Entomophaga* 8: 183–190

Legaspi JC and French JV (1996) The citrus leafminer and its natural enemies. Circ. B96-1, Texas A & M – Kingsville Citrus Center, Weslaco, Texas. Circular B96-1

LeRoux E (1971) Biological control attempts on pome fruits (apple and pear) in North America. 1860–1970. *The Canadian Entomologist* 103: 963–974

Lima M and Araujo J (1996) Egg parasitoid of *Phoracantha semipunctata* F. (Coleoptera: Cerambycidae) in southern Portugal. 1 - Field studies on the efficiency of *Avetainella longoi* Siscaro (Hymenoptera: Encyrtidae). In: International Congress of Entomology (ed.) *XX International Congress of Entomology : proceedings : Firenze, Italy, August 25–31, 1996*. The Congress, Firenze, Italy

Lipa JJ (1976) Gatunki stawonogów (Arthropoda) introdukowane do pokski w lutach 1959–1974 pzez instytut ochrony roslin celem biologiznego zwalczania szkoduików. *Prace Naukowe Instytutu Ochrony Roslin* 18: 157–166

Llácer E, Urbaneja A, Garrido A and Jacasa J-A (2005) Morphology and development of immature stages of *Galeopsomyia fausta* (Hymenoptera: Eulophidae: Tetrastichinae). *Annals of the Entomological Society of America* 98: 747–753

Llácer E, Urbaneja A, Garrido A and Jacas J (2006) Temperature requirements may explain why the introduced parasitoid *Quadrastichus citrella* failed to control *Phyllocnistis citrella* in Spain. BioControl 51: 439–452

Lloyd DC (1972) Some North American parasites of the potato tuber moth *Phthorimaea operculella* (Zeller) and remarks on those in other continents. *Technical Bulletin of the Commonwealth Institute for Biological Control* 15: 35–49

Loomans AJM (2007) Regulation of invertebrate Biological Control Agents in Europe: review and recommendations in its pursuit of a harmonised regulatory system. Report of EU project REBECA [Regulation of Biological Control Agents]. Available at: www.rebeca-net.de/downloads/Regulation%20of%20Beneficials%20in%20Europe.pdf (accessed 25 May 2016)

Longo S (1984) Distribution and density of scale insects (Homoptera, Coccoidea) on olive trees in Eastern Sicily. In: Cavalloro R and Crovetti A (eds) *Proceeding of the CEC / FAO / IOBC International Joint Meeting - Integrated Pest Control in Olive Groves, Pisa, Italy*. A.A. Balkema, Rotterdam, the Netherlands, pp. 159–168

Longo S and Siscaro G (1997) Recent approaches to the biological control of the citrus leafminer in Italy. *Bulletin OILB/SROP* 20: 71–75

Louda SM, Pemberton RW, Johnson MT and Follett PA (2003) Nontarget effects: the Achilles' heel of biological control? *Annual Review of Entomology* 48: 365–396

Lucchi A, Santini L and Wilson S (2003) Biological control of an introduced flatid planthopper by *Neodryinus typhlocybae* (Ashmead) (Hymenoptera: Dryinidae) in Europe. Poster presented at the ESA Annual Meeting and Exhibition, 2003 in Cincinnati, Ohio. Available at: http://esa.confex.com/esa/2003/techprogram/paper_12241.htm (accessed 5 May 2016)

Luck RF (1981) Parasitic insects introduced as biological control agents for arthropod pests. In: Pimentel D (ed.) *CRC Handbook of Pest Management in Agriculture*. CRC Press Inc, Boca Raton, Florida, pp. 125–284

Lynch LD, Hokkanen HMT, Babendreier D, Bigler F, Burgio G, *et al.* (2001) Insect biological control and non-target effects: a European perspective. In: Wajnberg E, Scott JK and Quimby PC (eds) *Evaluating Indirect Ecological Effects of Biological Control.* CAB International, Wallingford, UK, pp. 99–125

Maceljski M and Igrc J (1989) The phytophagous insect fauna of *Ambrosia artemisiifolia* in Yugoslavia. In: Delfosse ES (ed.) *Proceedings of the VII International Symposium on Biological Control of Weeds*. Istituto sperimentale per la patologia vegetale di roma, Ministero dell' Agricoltura e delle Foreste, Rome, pp. 639–643

McMurtry JA, Johnson HG and Newberger SJ (1992) Imported parasite of greenhouse thrips established on California avocado. *California Agriculture* Nov-Dec: 31–32

McPartland JM, Clarke RC and Watson DP (2000) *Hemp Diseases and Pests: Management and Biological Control.* CAB International, Wallingford, UK

Maini S, Nicoli G and Manzaroli G (1990) Evaluation of the egg parasitoid *Endovum puttleri* Grissell (Hym. Eulophidae) for the biological control of *Leptinotarsa decemlineata* (Say) (Col. Chrysomelidae) on eggplant. *Bolletino dell' Istituto di Entomologia 'Guido Grandi'* 44: 161–168

Malausa JC (1998) Un espoir face aux pullulations de *Metcalfia pruinosa.* Introduction en France de *Neodryinus typhlocybae*, parasite larvaire de cette "cicadelle". *Phytoma* 512: 37–40

Malausa JC and Girardet N (1997) Lutte biologique contre le psylle de l'eucalyptus. Acclimatation sur Cote d'Azur d'un auxiliaire prometteur, *Psyllaephagus pilosus. Phytoma* 50: 49–51

Malausa JC, Franco E and Brun P (1988) Acclimatation sur la Côte d'Azur et en Corse de *Serangium parcesetosum* (Col.: Coccinellidae) prédateur de l'Aleurode des Citrus, *Dialeurodes citri* (Hom. : Aleyrodidae). *Entomophaga* 33: 517–519

Malausa JC, Rabasse J-M and Kreiter P (2008) Les insectes entomophages d'intérêt agricole acclimatés en France métropolitaine depuis le début du 20 ème siècle. *Bulletin OEPP/EPPO* 38: 136–146

Manojlovic B, Zabel A, Peric P, Stankovic S, Rajkovic S and Kostic M (2003) *Dendrosoter protuberans* (Hymenoptera: Braconidae), an important elm bark beetle parasitoid. *Biocontrol Science and Technology* 13: 429–439

Mansilla JP and Pérez Otero R (1996) El defoliador del eucalipto *Gonipterus scutellatus. Phytoma España* 81: 36–42

Marchal P (1913) L'*Icerya purchasi* en France et l'acclimatation de son ennemi d'origine australienne, le *Novius cardinalis. Annales des Epiphyties* 1: 13–26

Marchal P (1929) Les ennemis du puceron lanigère, conditions biologiques et cosmiques de sa multiplication - Traitements. *Annales des Epiphyties* 15: 125–181

Mathys G and Guignard E (1965) Etude de l'efficacité de *Prospaltella perniciosi* Tow. en Suisse, parasite du pou de San-José. *Entomophaga* 10: 193–220

Mazzone P and Viggiani G (1985) Nuovi dati sull'*Encarsia pergandiella* (Howard) (Hym.: Aphelinidae), parassita esotico introdotto in Italia contro il *Trialeurodes vaporarium* (Westw.) (Hom.: Aleyrodidae). Atti XIV Congresso Nazionale Italiano di Entomologia, Palermo, Italy, pp. 855–859

Menzel R (1931) Beobachtungen über das diesjährige Verhalten des Blutlausparasiten *Aphelinus mali. Schweizerische Zeitschrift für Obst- und Weinbau* 5-6: 117–119

Michelakis SE (1997) The citrus leafminer in Greece. *Bulletin OILB/SROP* 20: 81–82

Mifsud D (1997) Biological control in the Maltese Islands – past initiatives and future programmes. *EPPO Bulletin* 27: 77–84

Mills N, Pickel C, Mansfield S, Mcdougall S, Buchner R, *et al.* (2000) Mass releases of *Trichogramma* wasps can reduce damage from codling moth. *California Agriculture* 56: 22–25

Minkenberg OPJM (1989) Temperature effects on the life history of the eulophid wasp *Diglyphus isaea*, an ectoparasitoid of leafminers (*Liriomyza* spp.), on tomatoes. *Annals of Applied Biology* 115: 381–397

Moens R (1963) Essai d'acclimatation de *Perillus bioculatus* Fabr., prédateur du doryphore. *Mededelingen van de Landbouw Hogeschool Gent* 28: 792–810

Monastero S (1960) Una eccezionale cattura di *Opius siculus* Mon. nell'annata agraria 1958. *Bollettino dell'Istituto di Entomologia Agraria, Universita di Palermo* 3: 261–267

Monastero S and Genduso P (1962) La lotta biologica contro la mosca delle olive. Possibilità di allevamento e diffusione degli *Opius* trovati in Sicilia. *Bollettino dell'Istituto di Entomologia Agraria, Universita di Palermo* 4: 31–51

Mouna M and Fabre JP (2005) Pest insects of cedars: *Cedrus atlantica* Manetti, *C. libani* A. Richard and *C. brevifolia* Henry in the Mediterranean area. In: Lieutier F and Ghaioule D (eds) *Entomological Research in Mediterranean Forest Ecosystems.* INRA, Paris, pp. 89–104

Mozaffarian F, Mardi M, Sarafrazi A and Ganbalani GN (2007) Host plant-associated population variation in the carob moth *Ectomyelois ceratoniae* in Iran: a geometric morphometric analysis suggests a nutritional basis. *Journal of Insect Science* 8: 1–11

Myartseva SN and Coronado-Blanco JM (2007) Especies de *Eretmocerus* Haldeman (Hymenoptera: Aphelinidae) – parasitoides de *Aleurothrixus floccosus* (Maskell) (Homoptera: Aleyrodidae) de México, con la descripción de una nueva especie. *Acta Zoologica Mexicana (Nueva Serie)* 23: 37–46

Nadel H, Daane KM, Hoelmer KA, Pickett CH and Johnson MW (2009) Non-target host risk assessment of the idiobiont parasitoid *Bracon celer* (Hymenoptera: Braconidae) for biological control of olive fruit fly in California. *Biocontrol Science and Technology* 19: 701–715

Nardi F, Carapelli A, Dallai R, Roderick GK and Frati F (2005) Population structure and colonization history of the olive fly, *Bactrocera oleae* (Diptera, Tephritidae). *Molecular Ecology* 14: 2729–2738

Nechols JR, Andres LA, Beardsley JW, Goeden RD and Jackson CG (1995) *Biological Control in the Western United States: Accomplishments and Benefits of Regional Research Project* W-84, 1964–189. University of California, Division of Agriculture and Natural Resources Oakland, California

Neuenschwander P (1982) Searching parasitoids of *Dacus oleae* (Gmel) (Dipt, Tephritidae) in South Africa. *Journal of Applied Entomology* 94: 509–522

Nikbakhtzadeh ST (2002) Blister beetles (Coleoptera: Meloidae) in Nahavand County (Hamedan Province, Iran) and their ecological relationship to other Coleopteran families. *Iranian Journal of Public Health* 31: 55–62

Nonell Comas J (1938) Introducción y difusión del *Aphelinus mali* (Hald.) en España. *Boletín de* Patologia *Vegetal y Entomología Agricola* 8: 179–186

Noyes JS (2002) Interactive catalogue of world Chalcioidea 2001. Compact disc. Taxpad, Vancouver, Canada

Noyes JS (2016) Universal Chalcidoidea Database. World Wide Web electronic publication. Available at: http://www.nhm.ac.uk/chalcidoids (accessed May 2016)

OECD (2003) Guidance for information requirements for regulations of invertebrates as biological control agents. OECD Environment, Health and Safety Publications. Series on Pesticides 21, 22 pp.

O'Hara JE (2005) A review of the tachinid parasitoids (Diptera: Tachinidae) of Nearctic *Choristoneura* species (Lepidoptera: Tortricidae), with keys to adults and puparia. *Zootaxa* 938: 1–46

O'Hara JE and Mahony SJ (2009a) Tachinidae Resources; Genus *CEROMASIA* Rondani, 1856. Available at: http://www.nadsdiptera.org/Tach/Nearctic/CatNAmer/Genera/Ceromasia.html (accessed 8 June 2016)

O'Hara JE and Mahony SJ (2009b) Tachinidae Resources; Genus *MYIOPHARUS* Brauer and Bergenstamm, 1889. Available at: http://www.nadsdiptera.org/Tach/Nearctic/CatNAmer/Genera/Myiopharus.html (accessed 8 June 2016)

O'Hara JE and Mahony SJ (2009c) Tachinidae Resources; Genus *SMIDTIA* Robineau-Desvoidy, 1830. Available at: http://www.nadsdiptera.org/Tach/Nearctic/CatNAmer/Genera/Smidtia.html (accessed 8 June 2016)

Oboyski PT (2009) Parasitoids (Ichneumonidae, Eulophidae) of *Cydia* moth larvae in Hawaii. Available at: http://nature.berkeley.edu/~poboyski/Lepidoptera/parasitoids.htm (accessed 5 May 2016)

Obrycki JJ and Tauber M (1982) Thermal requirements for development of *Hippodamia convergens* (Coleoptera: Coccinellidae). *Annals of the Entomological Society of America* 75: 678–683

Obrycki JJ, Tauber MJ, Tauber CA and Gollands B (1985) *Edovum puttleri* (Hymenoptera: Eulophidae), an exotic egg parasitoid of the Colorado potato beetle (Coleoptera: Chrysomelidae): responses to temperate zone conditions and resistant potato plants. *Environmental Entomology* 14: 48–54

OILB (Organisation Internationale de Lutte Biologique) (1970) Groupe de travail: Cochinelles des agrumes. Réunion tenue au Maroc (26 au 31 Octobre 1970). *Al Awamia* 37: 109 pp.

Onillon JC (1973) Possibilités de régulation des populations d'*Aleurothrixus floccosus* Mask. (Homopt., Aleurodidae) sur agrumes par *Cales noacki* How. (Hymenopt., Aphelinidae). Bulletin of the EPPO 3: 17–26

Onillon JC and Onillon J (1972) Contribution à l'étude de la dynamique des populations d'homoptères inféodés aux agrumes. III. Introduction dans les Alpes-Maritimes de *Cales noacki* How. (Hym. Aphelinidae), parasite d'*Aleurothrixus floccosus* Mask. (Hom. Aleurodidae). Comptes Rendus des Séances de I'Academie d'Agriculture de France 58: 365–370

Onillon JC and Onillon J (1974) Contribution à l'étude de la dynamique des populations d'homoptères inféodés aux agrumes. III-2. Modalités de la dispersion de *Cales noacki* How. (Hym. Aphelinidae), parasite d'*Aleurothrixus floccosus* Mask. (Hom. Aleurodidae). *Bulletin WPRS* 1974/3: 51–56

OPIE (1986) Lutte biologique à l'aide d'insectes entomophages. Cahiers de Liaison Office pour l'Information Eco-entomologique (OPIE), 20, 61, 48 pp.

Orphanides GM (1984) Competitive displacement between *Aphytis* spp. (Hym. Aphelinidae) parasites of the California red scale in cyprus. *BioControl* 29: 275–281

Orphanides GM (1993) Control of *Saissetia oleae* (Hom.: Coccidae) in Cyprus through establishment of *Metaphycus bartletti* and *M. helvolus* (Hym.: Encyrtidae). *Entomophaga* 38: 235–239

Orphanides GM (1996) Establishment of *Comperiella bifasciata* (Hym.: Encyrtidae) on *Aonidiella aurantii* (Hom.: Diaspididae) in Cyprus. *Entomophaga* 41: 53–57

Ortu S and Floris I (1989) Preliminary study on the control of *Phthorimaea operculella* (Zeller) (Lepidoptera: Gelechiidae) on potato crops in Sardinia. *Difesa delle Piante* 12(1-2): 81–88

Ortu S and Prota R (1986) Brevi considerazioni sulle recenti introdozioni in Sardegna di entomophagi a protezione della coltura agrumicola. *Frustula Entomologica* 7-8: 115–123

Ouvrard D (2010) Psyl'list – The World Psylloidea Database. Available at: http://www.hemiptera-databases.com/psyllist/ (accessed May 2016)

Ovruski S, Schliserman P, De Colli O, Peòaloza C, Oroòo L and Colin C (2006) The establishment of *Aceratoneuromyia indica* in three biogeographical regions of Argentina. *Florida Entomologist* 89: 270–273

Page AR and Lacey KL (2006) *Economic Impact Assessment of Australian Weed Biological Control.* CRC Weed Management, Technical Series No 10, p. 164

Pappas S and Viggiani G (1979) Introdotta a Corfu la *Prospaltella lahorensis* How. (Hym. Aphelinidae) parasita del *Dialeurodes citri* (Ashm.) (Hom. Aleyrodidae). *Bolletino de Laboratorio Entomologia Agraria, Portici* 36: 38–41

Paraskakis M, Neuenschwander P and Michelakis S (1980) *Saissetia oleae* (Oliv.) (Hom., Coccidae) and its parasites on olive trees in Crete, Greece. *Zeitschrift für Angewandte Entomologie* 90: 450–464

Parkes GT and Walter GH (2001) Mating behaviour and alternative oviposition sites for male eggs in the heteronomous hyperparasitoid *Coccophagus gurneyi* Compere (Hymenoptera: Aphelinidae). *Australian Journal of Entomology* 40: 74–78

Pedata PA, Hunter MS, Godfray HCJ and Viggiani G (1995) The population dynamics of the white peach scale and its parasitoids in a mulberry orchard in Campania, Italy. *Bulletin of Entomological Research* 85: 531–539

Pelekassis CD (1974) Historical review of biological control of citrus scale insects in Greece. *Bulletin WPRS* 1974/3: 14–19

Perera PACR (1977) A technique for laboratory mass-breeding of *Eriborus trochanteratus* (Hym.: Ichneumonidae) a parasite of the coconut caterpillar, *Nephantis serinopa* (Lep.: Xylorictidae). *BioControl* 22: 217–221

Pina T and Verdú MJ (2007) Establecimiento y dispersión de *Aphytis melinus* y *A. lingnanensis* (Hym.: Aphelinidae), dos parasitoides introducidos para el control de *Chrysomphalus dictyospermi* Morgan y *Aonidiella aurantii* (Maskell) (Hem.: Diaspididae) en cítricos de la Comunidad Valenciana. *Boletín de Sanidad Vegetal, Plagas* 33: 311–320

Pina T, Martinez B and Verdú MJ (2003) Field parasitoids of *Aonidiella aurantii* (Homoptera: Diaspididae) in Valencia (Spain). *IOBC WPRS Bulletin* 26: 109–115

Pinto JD, Oatman ER and Platner GR (1986) *Trichogramma pretiosum* and a new cryptic species occurring sympatrically in southwestern North America (Hymenoptera: Trichogrammatidae). *Annals of the Entomological Society of America* 79: 1019–1028

Polaszek A, Abd-Rabou S and Huang J (1999) The Egyptian species of *Encarsia* (Hymenoptera: Aphelinidae) (1999) a preliminary review. *Zoologische Mededelingen Leiden* 73: 131–163. Available at: http://www. repository.naturalis.nl/document/43551 (accessed August 2010)

Poorani J (2004) An annotated checklist of the Coccinellidae (Coleoptera) of Indian Subregion. Available at: http://www.angelfire.com/bug2/j_poorani/checklist.pdf (accessed 5 May 2016)

Principi MM, Domenichini G and Martelli M (1974) Lotta integrata e lotta guidata nei frutteti dell'Italia settentrionale. *Relazione 10. Congresso nazionale italiano di Entomologia, Sassari* 1974: 47 pp

Priore R (1971) Nota sull'effizienza biologica della *Rodolia cardinalis* Muls. e sulla sua diffusione in Italia. Note divulgazione dell' *Istituto di Entomologia* Agraria *dell Università di Napoli, Portici* 3: 3–5

Protasov A, Blumberg D, Brand D, La Salle J and Mendel Z (2007) Biological control of the eucalyptus gall wasp *Ophelimus maskelli* (Ashmead): taxonomy and biology of the parasitoid species *Closterocerus chamaeleon* (Girault), with information on its establishment in Israel. *Biological Control* 42, 196–206

Pruszynski S (1989) 100 lat introdukcji owadow pozytecznych w ochronie roslin. *Materialy XXIX sesji naukowej institutu ochrony roslin, Poznan*: 103–112

Pucci C, Spanedda AF and Minutoli E (2003) Field study of parasitism caused by endemic parasitoids and by the exotic parasitoid *Copidosoma koehleri* on *Phthorimaea operculella* in Central Italy. *Bulletin of Insectology* 56: 221–224

Pussard, R (1950) L'utilisation des insectes auxilliaires en France. *Bulletin technique d'Information des ingénieurs des services agricoles* 49: 5 pp.

Quacchia A, Moriya S, Bosio G, Scapin I and Alma A (2008) Rearing, release and settlement prospect in Italy of *Torymus sinensis*, the biological control agent of the chestnut gall wasp *Dryocosmus kuriphilus*. *BioControl* 53: 829–839

Rabitsch W (2010) True bugs (Hemiptera, Heteroptera). Chapter 9.1. In: Roques A, Kenis M, Lees D, Lopez-Vaamonde C, Rabitsch W, Rasplus J-Y and Roy D (eds) *Alien Terrestrial Arthropods of Europe. BioRisk* 4(1): 407–433

Rasplus J-Y, Villemant C, Paiva MR, Delvare G and Rasplus A (2010) Hymenoptera. Chapter 12. In: Roques A, Kenis M, Lees D, Lopez-Vaamonde C, Rabitsch W, Rasplus J-Y and Roy D (eds) *Alien Terrestrial Arthropods of Europe. BioRisk* 4(2): 669–776

Raworth DA, Gillespie DR and Shipp JL (2002) *Trialeurodes vaporariorum* (Westwood), greenhouse whitefly, and *Bemisia tabaci* (Gennadius), sweet potato whitefly. In: Mason PG and Huber JT (eds) *Biological Control Programmes in Canada, 1981–2000*. CAB International, Wallingford, UK, pp. 265–269

Reina P and La Salle J (2004) Two new species of *Quadrastichus* Girault (Hymenoptera: Eulophidae): parasitoids of the leafminers *Phyllocnistis citrella* Stainton (Lepidoptera: Gracillariidae) and *Liriomyza trifolii* (Burgess) (Diptera:Agromyzidae). *Journal of Hymenopteran Research* 13: 108–119

Rice SK, Westerman B and Federici R (2004) Impacts of the exotic, nitrogen-fixing black locust (*Robinia pseudoacacia*) on nitrogen cycling in a pine-oak ecosystem. *Plant Ecology* 174: 97–107

Robinson A and Hooper G (1989) *Fruit Flies: Their Biology, Natural Enemies and Control.* Elsevier, Amsterdam

Rondon SI (2010) The potato tuberworm: a literature review of its biology, ecology, and control. *American Journal of Potato Research* 87: 149–166

Roy H and Migeon A (2010) Ladybeetles (*Coccinellidae*). Chapter 8.4. In: Roques A, Kenis M, Lees D, Lopez-Vaamonde C, Rabitsch W, Rasplus J-Y and Roy D (eds) *Alien Terrestrial Arthropods of Europe. BioRisk* 4(1): 293–313

Roy H, Adriaens T, Isaac NJB, Kenis M, Onkelinx T, *et al.* (2012) Invasive alien predator causes rapid declines of native European ladybirds. *Diversity and Distributions*, doi:10.1111/j.1472-4642.2012.00883.x

Ruiz Castro A (1965) *Plagas y enfermedades de la vid.* Instituto Nacional de Investigaciones Agronomicas, Madrid, 757 pp.

Russo G (1951) La lotta biologica contro i parassiti animali dei vegetali in Italia. *Bollettino del Laboratorio di Entomologia Agraria Filippo Silvestri* 10: 86–97

Russo G (1958) Problemi italiani di lotta biologica dei fitofagi. *Bollettino del Laboratorio di Entomologia Agraria Filippo Silvestri* 16: 141–147

Russo G (1959) Experimenti di lotta biologica contro la mosca delle olive (*Dacus oleaea*) e la mosca della frutta (*Ceratitis capitata*) in Italia. *Bollettino del Laboratorio di Entomologia Agraria Filippo Silvestri* 17: 131–142

Sachtleben H (1929) Die Forleule –*Panolis flammea* Schiff. Julius Springer Verlag, Berlin

Sachtleben H (1941) Biologischen Bekämpfungsmassnahmen. In: Sorauer P (ed.) *Handbuch für Pflanzenkrankheiten 6.* Paul Parey, Berlin, pp. 1–120

Sankaran T and Girling DJ (1980) The current status of biological control of the potato tuber moth. *Biocontrol News and Information* 1: 207–211

Sarýkaya O and Avcý M (2006) The occurence of *Choristoneura murinana* in *Abies cilicia* forests in Asia minor. *Munis Entomology and Zoology* 1(2): 289–296

Sasso R, Laudonia S and Viggiani G (2008) Dati preliminari sul controllo biologico di *Ophelimus maskelli* (Ashmead) (Hymenoptera: Eulophidae) in Campania al seguito dell'introduzione del suo antagonista *Closterocerus chamaeleon* (Hymenoptera: Eulophidae). *Bollettino del Laboratorio di Entomologia Agraria Filippo Silvestri* 62: 51–55

Saucke H, Juergens M, Döring TF, Fittje S, Lesemann D-E and Vetten H-J (2008) Einfluss von Saatzeit und Strohmulch auf Blattlausbesiedelung und Viruserkrankungen in Fababohnen (*Vicia faba* L.). *Mitteilungen der Deutschen Gesllschaft für allgemeine und angewandte Entomologie* 16: 339–342

Schauff ME, La Salle J and Wijesekara GA (1998) The genera of chalcid parasitoids (Hymenoptera: Chalcidoidea) of citrus leafminer *Phyllocnistis citrella* Stainton (Lepidoptera: Gracillaridae). *Journal of Natural History* 32: 1001–1056

Schaufuß C and Schenkung S (1910) *Deutsche Entomologische Rundschau im Gebiete der Insektenkunde mit besonderer Berücksichtigung der Literatur.* Deutschen Entomologischen National-Museum, Berlin

Scheffer SJ and Lewis ML (2006) Mitochondrial phylogeography of the vegetable pest *Liriomyza trifolii* (Diptera: Agromyzidae): diverged clades and invasive populations. *Annals of the Entomological Society of America* 99: 991–998

Schoevers, TAC (1934) De invoer, vetiging en verbreiding van het sluipwespje *Aphelinus mali* Say, parsiet van de bloedluis in Nederland. *Tijdschrift over Plantenziekten* 40: 273–278

Sears MK and Boiteau G (1989) Parasitism of Colorado potato beetle (Coleoptera, Chrysomelidae) eggs by *Edovum puttleri* (Hymenoptera, Eulophidae) on potato in eastern Canada. *Journal of Economic Entomology* 82: 803–810

Segoli M, Bouskila A, Harari AR and Keasar T (2009) Developmental patterns in the polyembryonic parasitoid wasp *Copidosoma koehleri*. *Arthropod Structure and Development* 38: 84–90

Seiger L (1991) Element stewardship abstract for *Polygonum cuspidatum*. The Nature Conservancy. Available at: www.invasive.org/weedcd/pdfs/tncweeds/polycus.pdf (accessed 26 May 2016)

Shaw RH, Bryner S and Tanner R (2009) The life history and host range of the Japanese knotweed psyllid, *Aphalara itadori* Shinji: potentially the first classical biological weed control agent for the European Union. *Biological Control* 49: 105–111

Shaw RH, Tanner R, Djeddour D and Cortat G (2011) Classical biological control of *Fallopia japonica* in the United Kingdom – lessons for Europe. *Weed Research* 51: 552–558

Silvestri F (1914) Report of an expedition to Africa in search of the natural enemies of fruit flies (Trypaneidae) with descriptions, observations and biological notes. *Bulletin of the Division of Entomology Board of Agriculture for Hawaii* 3: 176 pp.

Silvestri F (1926) Lotta contro alcune cocciniglie degli agrumi. *Nuovi annali dell' Agricoltura* 6: 97–101

Silvestri F (1938) La lotta biologica control le mosche dei fruittiferi della famiglia Trypetidae. *Verhandlungen des 7. Internationalen Kongresses für Entomologie* 1938 4: 2396–2418

Silvestri F (1945) Descrizione a biologia del coleottero stafilinide *Belonuchus formosus* Grav. introdotto in Italia per la lotta contra ditteri tripaneidi. *Bollettino del Laboratorio di Entomologia Agraria Filippo Silvestri* 5: 312–316

Sime KR, Daane KM, Andrews JW, Hoelmer K, Pickett CH, *et al.* (2006) The biology of *Bracon celer* as a parasitoid of the olive fruit fly. *BioControl* 51: 553–567

Simov N, Langourov M, Grozeva S and Gradinarov D (2012) New and interesting records of alien and native true bugs (Hemiptera: Heteroptera) in Bulgaria. *Acta zoologica bulgarica* 64(3): 241–252

Sinacori A, Mineo G and Lo Verde G (1993) Biological control of *Parabemisia myricae* (Kuwana) in western Sicily (Hom. Aleyrodidae). *IOBC WPRS Bulletin* 16(7): 41–44

Siscaro G, Barbagallo S, Longo S, Reina P and Zapala L (1999) Esiti dell'introduzione di parasitoidi esotici di *Phyllocnistis citrella* Stainton (Lepidoptera, Gracillaridae) in Sicilia. *Phytophaga (Palermo)* 9 (supplemento): 31–39

Siscaro G, Caleca V, Reina P, Rizzo MC and Zappalà L (2003) Current status of the biological control of the citrus leafminer in Sicily. *IOBC WPRS Bulletin* 26(6): 29–36

Spain (1941) Memoria de los trabajos realizados por la Estación de Fitopatologia agricola de la Coruña, años 1939–1940. *Publicaciones del Estación de Fitopatología Agrícola de La Coruña* 23: 7–19

Starý P (1976) *Aphid Parasites (Hymenoptera, Aphidiidae) of the Mediterranean Area*. Dr W Junk bv Publishers, The Hague, Netherlands

Starý P (1995) Natural enemy spectrum of *Aphis spiraephaga* (Hom.: Aphididae), an exotic immigrant aphid in Central Europe. *Entomophaga* 40: 29–34

Starý P (1999) Parasitoids and biocontrol of Russian wheat aphid, *Diuraphis noxia* (Kurdj.) expanding in central Europe. *Journal of Applied Entomology* 123: 273–279

Starý P (2002) Field establishment of *Aphidius colemani* Vier. (Hym., Braconidae, Aphidiinae) in the Czech Republic. *Journal of Applied Entomology* 126: 405–408

Starý P and Havelka J (2008) Fauna and associations of aphid parasitoids in an up-dated farmland area (Czech Republic). *Bulletin of Insectology* 61: 251–276

Starý P, Lyon JP and Leclant F (1988) Biocontrol of aphids by the introduced *Lysiphlebus testaceipes* (Cress.) (Hym., Aphidiidae) in Mediterranean France. *Journal of Applied Entomology* 105: 74–87

Stavraki HG (1985) Use of *Trichogramma* spp. against the carpophagous generation of *Prays oleae* (Bern.) in Greece. In: Cavalloro R and Crovelli A (eds) *Integrated Pest Control in Olive-groves*. Balkema, Rotterdam, Netherlands, pp. 242–246

Stein JD (1981) Notes on the biology of western spruce budworm, *Choristoneura occidentalis* (Lepidoptera: Tortricidae) in north central Washington. *Pacific Insects* 23: 101–106

Stouthamer R, Jochemsena P, Platner GR and Pinto JD (2000) Crossing incompatibility between *Trichogramma minutum* and *T. platneri* (Hymenoptera: Trichogrammatidae): implications for application in biological control. *Environmental Entomology* 29: 832–837

Strauss G (2009) Host range testing of the nearctic beneficial parasitoid *Neodryinus typhlocybae*. *BioControl* 54: 163–171

Świerczewski D and Stroiński A (2011) The first records of the Nearctic treehopper *Stictocephala bisonia* in Poland (Hemiptera: Cicadomorpha: Membracidae) with some comments on this potential pest. *Polish Journal of Entomology* 80: 13–22

Tadić MD (1958) Biological control of the fall webworm (*Hyphantria cunea* Dr.). *European Proceedings of 10 International Congress of Entomolgy, Montreal* 1956, 4: 855–858

Tadić MD (1962) Rapport numérique entre *Anastatus disparis* R. et *Ooencyrtus kuwanae* How. dans certaines localités de Yugoslavie. *Agronomiski glasnik* 548–552

Tadić M (1967) Odomacivanje Prospaltella perniciosi Tow. u Jugoslaviji. *Zaštita bilja* 93–97: 147–154

Tadić M and Bincev B (1959) *Ooencyrtus kuwanai* How. u Jugoslaviji. *Zaštita bilja* 52–53: 51–59

Templado J (1957) Datos sobre *Doencyrtus kuwanai* How. (Calcédido parásito do *Lymantria dispar* L.) en España. *Publicaciones del Instituto de Biologla* aplicada 25: 119–129

Tena A, Soto A and Garcia-Marí F (2008) Parasitoid complex of black scale *Saissetia oleae* on citrus and olives: parasitoid species composition and seasonal trend. *BioControl* 53(3): 473–487

Thomas DB (1994) Taxonomic synopsis of the Old World asopine genera (Heteroptera: Pentatomidae). *Insecta Mundi* 8: 145–212

Thompson WR (1930) *Biological Control of Insect and Plant Pests*. HMSO, London, 124 pp.

Urquijo P (1945) Ensayo del parásito útil *Trichogramma minutum* en la lucha contra la polilla de la patata. *Boletín de Patologia Vegetal y Entomología Agricola* 13: 281–300

Urquijo P (1951) Aplicación de la genética al aumento de la eficacia del *Trichogramma minutum* en la luch biológica. *Boletín de Patologia Vegetal y Entomología Agricola* 18: 1–12

Vajna L (2002) Downy mildew epidemic on common ragweed in Hungary caused by *Plasmopara halstedii*. *Plant Pathology* 51: 809

Vajna L, Bohár G and Kiss L (2000) First report of *Phyllachora ambrosiae* in Europe causing epidemics on common ragweed. *Plant Disease* 84: 489

van Achterberg C (1999) The Palaearctic species of the genus *Diachasmimorpha* Viereck (Hymenoptera: Braconidae: Opiinae). *Zoologiske Mededelingen Leiden* 73: 1–10

Van Averbeke A and Grégoire JC (1995) Establishment and spread of *Rhizophagus grandis* Gyll (Coleoptera: Rhizophagidae) 6 years after release in the Forêt domaniale du Mézenc (France). *Annals of Forest Science* 52: 243–250

Van Driesche RG, Idoine K, Rose M and Bryan M (1998) Release, establishment and spread of Asian natural enemies of *Euonymus* scale (Homoptera: Diaspididae) in New England. *Florida Entomologist* 81: 1–9

Van Driesche RG, Carruthers RI, Center T, Hoddle MS, Hough-Goldstein J, *et al.* (2010) Classical biological control for the protection of natural ecosystems. *Biological Control* 54: S2–S33 van Lenteren JC (2012) The state of commercial augmentative biological control: plenty of natural enemies, but a frustrating lack of uptake. *BioControl* 57: 1–20

van Lenteren JC and Bueno VHP (2003) Augmentative biological control of arthropods in Latin America. *BioControl* 48: 123–139

van Lenteren JC, Bale J, Bigler F, Hokkanen HMT and Loomans AJM (2006) Assessing risks of releasing exotic biological control agents of arthropod pests. *Annual Review of Entomology* 51: 609–634

van Lenteren JC, Bigler F, Babendreier D and Loomans AJM (2008) *Harmonia axyridis*: an environmental risk assessment for Northwest Europe. *BioControl* 53: 37–54

Vayssière P (1953) Recherches sur la lutte biologique contre les ennemis des cultures. *Comptes Rendus de l'Assemblée générale de l'Union Internationale des Sciences Biologique, Nice* 1953, 12 pp.

Vercher R, Garcia Mari F, Costa Comelles J, Marzal C and Granda C (2000) Importacion y establecimiento de parasitos del minador de hojas de citricos *Phyllocnistis citrella* (Lepidoptera: Gracillaridae). *Boletin de Sanidad Vegetal, Plagas* 26(4): 577–591

Vercher R, Garcia-Mari F, Costa-Comelles J, Marzal C and Villalba M (2003) Biological control of the citrus leafminer *Phyllocnistis citrella* (Lepidoptera: Gracillaridae) in Spain: native parasitoids and establishment of *Citrostichus phyllocnistoides* (Hymenoptera: Eulophidae). *IOBC WPRS Bulletin* 26(6): 7–15

Vidano C and Meotto F (1968) Moltiplicazione e disseminazione di *Polynema striaticorne* GIRAULT (Hymenoptera Mimaridae). *Annali della Facolta di Scienze Agrarie della Universita degli Studi di Torino* 4: 297–316

Vidano C, Arzone A and Arno A (1987) Researches on natural enemies of viticolous Auchenorrhynca. In: Cavalloro R (ed.) *Integrated Pest Control in Viticulture*. Proceedings of a meeting of the EC Experts' Group, Portoferraio, 26–28 September, 1985, Amsterdam, pp. 97–101

Viggiani G (1978) Current state of biological control of olive scales. *Bolletino del Laboratorio di Entomologia Agraria 'Filippo Silvestri', Portici* 35: 30–38

Viggiani G and Mazzone P (1983) Risultati preliminari sulla introduzione in Italia di *Metaphycus bartletti* Ann. e Myn., parassita di *Saissetia oleae* (Oliv.). *Atti XII Congresso Nazionale Italiano di Entomologia, Roma*, II: 305

Vreysen MJB, Hendrichs J and Enkerlin JW (2006) The sterile insect technique as a component of sustainable area/wide integrated pest management of selected horticultural insect pests. *Journal of Fruit and Ornamental Plant Research* 14: 107–131

Wahl B (1914) Die biologische Methode der Bekämpfung von Pflanzenschädlingen. Verhandlungen der 4. Tagung der Österreichischen Obstbau- und Pomologie-Gesellschaft, Vienna 1914: 19 pp.

Wang L, Bisseleua DHB, You M, Huang J and Liu B (2006) Population dynamics and functional response of *Citrostichus phyllocnistoides* (Narayanan) (Hym., Eulophidae) on citrus leaf-miner, *Phyllocnistis citrella* Stainton (Lep., Phyllocnistidae) in Fuzhou region of south-east China. *Journal of Applied Entomology* 130: 96–102

Wang XG and Messing RH (2004) Two different life-history strategies determine the competitive outcome between *Dirhinus giffardii* (Chalcididae) and *Pachycrepoideus vindemmiae* (Pteromalidae), ectoparasitoids of cyclorrhaphous Diptera. *Bulletin of Entomlogical Research* 94: 473–480

WaspWeb (2004–2010) Wasps of Africa and Madagascar: Eulophidae – Classification of afrotropical eulophid wasps. Available at: http://www.waspweb.org/Chalcidoidea/Eulophidae/Classification/index.htm (accessed 5 May 2016)

Waterhouse DF (1993) *Biological Control Pacific Prospects – Supplement 2*. Australian Centre for International Agricultural Research Canberra

Waterhouse DF (1998) *Biological Control of Insect Pests: Southeast Asian Prospects*. Australian Centre for International Agricultural Research (ACIAR Consultant in Plant Protection), Canberra

Waterhouse DF and Sands DPA (2001) *Classical Biological Control of Arthropods in Australia*. CSIRO Entomology and CSIRO Publishing Canberra, Melbourne, Australia

Weaver CR (1949) Some aspects of the distribution of larval parasites of the oriental fruit moth in Ohio. *Ohio Journal of Science* 49: 154–159

Weber DC, Rowley DL, Greenstone MH and Athanas MH (2006) Prey preference and host suitability of the predatory and parasitoid carabid beetle, *Lebia grandis*, for several species of Leptinotarsa beetles. *Journal of Insect Science* 6: 1–14

Wharton RA and Yoder MJ (2016) Parasitoids of Fruit-Infesting Tephritidae. Available at: http://paroffit.org (accessed 25 May 2016)

Whitebread SE (1989) *Phyllonorycter robiniella* (Clemens. 1859) in Europe (Lepidoptera, Gracilariidae). *Nota Lepidopterologica* 12: 344–353

Willemstein SC (1987) *An Evolutionary Basis for Pollination Ecology*. Leiden University Press, Leiden, Netherlands

Wood BJ (1962) Natural enemies in the control of citrus pests in Cyprus. Pt.1. Making the most of natural enemies. Pt.2. The insectary built at Fassouri and a preliminary report on newly imported natural enemies. *Countryman Nicosia* 1962: 8-11 and 16–22

Wood BJ (1963) Imported and indigenous natural enemies of *Citrus* coccids and aphids in Cyprus, and an assessment of their potential value in integrated control programmes. *Entomophaga* 8: 67–82

Wylie HG (1977) Observations on *Mericia ampelus* (Dip: Tachinidae), an occasional parasite of Bertha armyworm, *Mamestra configurata* (Lep: Noctuidae), in Western Canada. *Canadien Entomologist* 109: 1023–1024

Yefremova Z (2008) Order Hymenoptera, family Eulophidae. *Arthropod Fauna of the UAE* 1: 345–360

Yu DS, van Achterberg K and Horstmann K (2005) World Ichneumonoidea 2004. Taxonomy, Biology, Morphology and Distribution. CD/DVD. Taxapad, Vancouver, Canada. Available at: www.taxapad.com (accessed 5 May 2016)

Zimmermann O (2004) Der Einsatz von *Trichogramma*-Schlupfwespen in Deutschland: Zum aktuellen Stand der Forschung und Nutzung von Eiparasitoiden gegen Schadlepidopteren im biologischen Pflanzen- und Vorratsschutz. *Gesunde Pflanzen* 56: 157–166

Zinna G (1960a) Ricerche sugli insett entomophagi. 1. Specialazione entomoparasitica negli Encyrtidae: studio morphologica etologica e fisiologica del *Leptomastix dactylopii* Howard. *Bollettino del Laboratorio di Entomologia Agraria Filippo Silvestri* 18: 1–148

Zinna G (1960b) Esperimente di lotta biologica contro i cotonello degli agrumi (*Pseudococcus citri* Risso) nell'isola Procida mediante i'impiego di due parassiti esotici *Pauridia peregrina* Timb. e *Leptomastix dactylopii* How. *Bollettino del Laboratorio di Entomologia Agraria Filippo Silvestri* 18: 257–284

Zubrik M and Novotny J (1997) Egg parasitization of *Lymantria dispar* (Lepidoptera, Lymantriidae) in Slovakia. *Biológia, Bratislava* 52(2): 343–350

Zwölfer H (1965) Preliminary list of phytophagous insects attacking wild Cynareae (Compositae) in Europe. *Technical Bulletin Commonwealth Institute of Biological Control* 6: 81–154

Index

Names in **bold** refer to target species. Where a Latin name is followed by another Latin name in brackets, the first name is a synonym of the name given in brackets.